THE BASICS OF

SELF-
BALANCING
PROCESSES

True Lean Continuous Flow

THE BASICS OF

SELF-BALANCING PROCESSES

True Lean Continuous Flow

Gordon Ghirann

CRC Press
Taylor & Francis Group
Boca Raton   London   New York

CRC Press is an imprint of the
Taylor & Francis Group, an **informa** business

A PRODUCTIVITY PRESS BOOK

CRC Press
Taylor & Francis Group
6000 Broken Sound Parkway NW, Suite 300
Boca Raton, FL 33487-2742

© 2012 by Taylor & Francis Group, LLC
CRC Press is an imprint of Taylor & Francis Group, an Informa business

No claim to original U.S. Government works

Printed in the United States of America on acid-free paper
Version Date: 20111227

International Standard Book Number: 978-1-4398-1965-4 (Paperback)

Visit the Taylor & Francis Web site at
http://www.taylorandfrancis.com

and the CRC Press Web site at
http://www.crcpress.com

To Jack, who gave me this gift.

To Holly, Isabella, Ava, and Gabriel,

who remind me every day what is possible.

Contents

Foreword

Anyone who has experimented with "Lean" production system application has, invariably, reached plateaus, or run into barriers that seem insurmountable. Fortunately, for those who support production operations, this book provides an approach that can help break through these barriers and get closer to the ideal of true continuous flow.

I make this claim with first-hand experience. As an internal consultant and operations manager, I had spent a few years implementing cellular production before I met Gordon Ghirann. I was fortunate to have his help implementing this "work distribution" method he calls *Self-Balancing*. Once I understood Self-Balancing, I could finally see the real barriers to flow, and I realized that these aforementioned "cells" were simply push systems in a smaller footprint. My eyes had been opened—I was seeing flow in a whole new way.

In this book, Gordon will challenge your mental-models of cellular manufacturing. Instead of the typical approach that seeks to balance the work and operate strictly according to takt time, Gordon describes a way to take advantage of the "human element" and release the full potential of the workers. Some readers will see this as a revolutionary approach that is ideal for their operations. Others more deeply rooted in the classic leveled approach may find themselves dismissing it. For those, keep in mind that the classic approach, which seeks to force leveling and takt-pacing, works well if you have (or have committed to developing) a complete lean management system that responds to problems every takt-time. The classic approach will continuously surface problems that must be solved. Couple this type of work system with a management system that can effectively problem-solve and you have the ingredients that make up Toyota's production system. However, since most organizations do not have the management capacity to respond every takt, Self-Balancing provides an extremely efficient (and effective) delivery method while they work to develop the management support systems.

That said, for years I have been telling people about how Self-Balancing surfaces barriers to flow and provides a framework for the supporting management support system. Invariably, they ask me where they can find out more. Until now, there was no source for this powerful concept outside of some limited research on the web.* Gordon, however, has probably experimented more with Self-Balancing than anyone else, so I'm very excited that he has finally brought this powerful concept to you. Moreover, as I read this book I was reminded of how most of what I know about continuous flow, I learned by working with Gordon and Self-Balancing. Now with this book, you will find that designing Self-Balancing production systems actually gives you new eyes for waste. Whether you are just getting started or have been practicing for some time, he makes the concept accessible for simple cases and also illustrates the important nuances that will help you apply it to very complex value streams. Gordon even includes advanced production system design elements, such as horizontal presentation and curtain/back-bench, that become necessary for Self-Balancing, but can work in any lean production system.

Use Self-Balancing as you problem-solve your way to true continuous flow. Use it to help you see your problems, and use it to engage the workers in solving these problems while you improve the performance of your organization's value streams to better serve customers.

Kirk Paluska
October 2011

---

* There is a body of research under the topic of "bucket-brigades" that shares many characteristics with Self-Balancing.

Acknowledgments

Given the span of time that has passed since I began formulating this book, I have many people to acknowledge.

First, my wife, Holly, who has always supported me in so many ways while I was developing and writing the material for this book. My three amazing kids for always being up for another Self-Balancing experiment: moving firewood, groceries, helping someone move, and on and on.

My parents for raising me to be practical, thoughtful, and efficient.

I have had many individuals supporting me directly with the work of the book: Didi, Jonnie, Dena, Ana, Caryl, and Ted; my editors, Jeanne Gearing (and typist), George Taninecz, and Kirsten Miller, and my illustrator Nataliya (Natasha) Shishkina; many coworkers, colleagues, and clients, Irving Leung, Kirk Paluska, Jim McMillin, Jeff, Jerry, Hung, Calvin, Bill, Bob, Doug, Larry, Terry, Trevor, Dave, Troy, Denise, Andrea, and Dina; and Landmark Education and the Wisdom Division .

And finally, my mentor, the person who taught me Self-Balancing and pretty much everything else I learned about manufacturing, Jack Zimmermann. His words and teachings are everywhere in this book.

The Author

Gordon Ghirann received his B.S. in Mechanical Engineering from Cal Poly, San Luis Obispo. He began developing the Self-Balancing methodology in 1998, working at and with several companies, products lines, and processes around the world, sharing, consulting, and teaching along the way. Gordon has been a regularly featured presenter of Workshops at the Association of Manufacturing Excellence (AME) conference. He currently resides in California. For more information visit www.self-balancing.com.

Introduction

What if there was a way for you to increase your manufacturing productivity, improve efficiency, and enhance employee satisfaction? Wouldn't you want to give it a try?

There is a way, and it's called Self-Balancing. I've developed this process and tested its limits over the course of more than 12 years. Over the course of those 12 years, I've refined Self-Balancing, but I've also looked for flaws. I've tried to break it and failed. I've taught, shared, consulted, and implemented Self-Balancing all around the world, with all kinds of products and industries: electronics, micro-optics, consumer products, and military products, to name a few. My clients' results have been off the charts. And so this book was born.

Although the book focuses primarily on manufacturing assembly lines, the methods apply far beyond the traditional manufacturing environment. And this book isn't just for industrial engineers: it's for line supervisors, process engineers, managers, product development engineers, and even your human resources department.

Examples of Self-Balancing have been around for probably thousands of years, as humans have worked together to move and assemble things. But since the days of the industrial revolution and subsequent assembly lines, industry has struggled to effectively create true continuous flow of products and services.

The current state of manufacturing and services, and specifically assembly lines, looks a lot like it has for the past 100 years. Although in some parts of the world, and at many companies, conditions for line workers have improved; in many other places, this isn't always the case. Sweat shops still exist, labor is exploited, and the line worker is relegated to monotonous, dehumanizing work.

Creating flow in the factory, large or small, still eludes us. There are chronic balance problems on assembly lines full of visible and hidden wastes. Traditional methods of setting up and balancing assembly lines are very engineering-intensive and somewhat cumbersome. Once the line is set up, it's fairly inflexible and not likely or easy to change.

THE SELF-BALANCED DIFFERENCE

If you're looking for a competitive advantage, Self-Balancing will provide just that. It'll not only improve working conditions but will create a breakthrough in the line worker's experience. Why try Self-Balancing? First, you can expect a productivity improvement greater than 30% *minimum*. Improvements of 50–60% percent and even up to 100% productivity are not uncommon. These improvements are possible because Self-Balancing is not just a tweak or change to assembly line balancing but a completely transformed method for achieving continuous flow. On the line, your operators will now function as a team, part of a dynamic, highly productive assembly line. They'll communicate and collaborate on each unit coming down the line, without the monotony of traditional assembly lines. Your workers will have the opportunity to improve and grow, and your best workers will finally be able to contribute according to their skills. Line supervisors will be freed up to focus on other tasks rather than directing the work of the line.

Here's what to expect from the book. For context, we'll first take a look at how the first assembly lines were operated in Henry Ford's Highland Park plant in Michigan. These methods have remained essentially unchanged for 100 years. Next, we'll look at the shortcomings of traditional assembly line balancing. If you've worked on, supervised, or designed assembly lines—or even if you've seen the "I Love Lucy" episode where Ethel and Lucy wrap chocolates on the conveyer belt—you'll realize that they have some limitations. After we delve into those, I'll start to walk you through the new paradigm of Self-Balancing. You'll learn how to deal with batch and off-line processes, how to debug your line, arrange your parts and tools, how to design a Self-Balanced cell, and much more.

Let's get started.

1

Conventional Assembly Line Balancing

Assembly lines have been around for nearly 100 years, dating back to Henry Ford's first moving assembly lines in Highland Park, Michigan, which produced Model Ts. Ford's assembly lines, along with his many other innovations, revolutionized manufacturing: flowing assembly through the plant, speeding production and lowering costs, and making high-volume manufacturing possible. Ford's assembly line concepts—single-piece flow, interchangeable parts, paced lines, and level loading, among others—have been used throughout the world and are still widely used today. Certainly, they are far more efficient than prior methods such as bay building.

Prior to Ford's assembly lines, the car, or boat, or whatever you were building, did not move. It stayed in one spot, a *bay*, while everything was brought to it. Since the car was not moving, there was no sense of urgency during the build process. The worker tended to be a highly skilled craftsman who had to know how to build the car start to finish. High-volume mass production was not possible because:

- It took a long time for one worker to build an entire car.
- There were not that many skilled workers to go around.
- To build many cars at once would require entire duplicate sets of tools, fixtures, jigs, etc., at each bay.

Let us take a look at the two types of assembly lines currently in use: paced lines (such as Henry Ford's first line) and manual lines.

PACED LINES

A cornerstone concept of assembly lines in the automotive industry and similar industries is a *paced line*—an assembly line in which the units being assembled are pulled or conveyed through the factory at a constant rate (see Figure 1.1).

As the term implies, in a paced line, a unit is in place at each assembly workstation for the same amount of time. Ford used this approach to balance the work content for each operator on the line, in an attempt to ensure that each operator at a workstation on the line was always working on a unit, and that all operators were working in unison until the next single unit was conveyed into their station. Ford's approach came to be known as "flow production," a moving assembly line where operators worked along the line on one unit at a time. Ford paced his moving line to make one-piece flow possible, structuring the work content to allow for approximately the same amount of work time per unit and per workstation.

MANUAL ASSEMBLY LINES

In a manual assembly line or cell, units are not connected by a cable or conveyor that moves them through the line. Instead, operators manually move units from one workstation to the other. Operators on manual assembly lines are disconnected from one another, in the sense that they are starting and finishing their tasks independently (see Figure 1.2).

Because operators cannot all work at the same exact pace and the amount of work, or *work content,* at each station is never exactly the same for all stations, a buffer of units—a small amount of work-in-process (WIP) inventory—is usually placed between each station. WIP inventory provides a buffer against work imbalance between stations.

This basic method for setting up an assembly line, or cell, both moving and manual-paced processes, has essentially been the same for the past 100 years. In an attempt to balance the work content for all of the operators, *level loading* was developed (see Figure 1.3).

Conveyor Moving Assembly Line

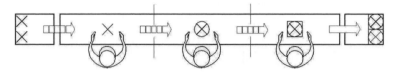

FIGURE 1.1
Paced line.

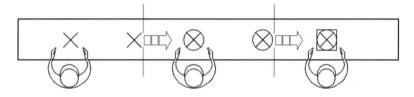

FIGURE 1.2
Manual assembly line.

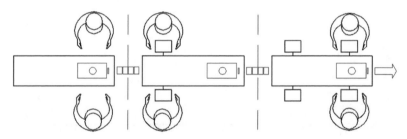

FIGURE 1.3
Automotive line, level loaded.

LEVEL LOADING

In level loading, all of the work content to be performed on an assembly line or cell is divided as evenly as possible among the operators. The goal is to have as little variation as possible in each operator's work content so that the operators can work at the same pace. This approach creates the highest possible productivity and minimizes the wait time for each operator (see Figure 1.4).

Level loading breaks up all of the steps required to build a unit into discrete, smaller tasks. An operator on the assembly line needs to know only one part of the overall assembly process, and he or she repeats that step throughout the shift. With additional cross-training, an operator can

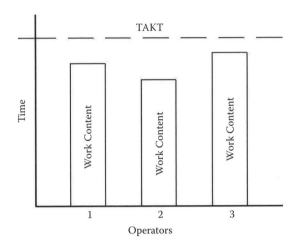

FIGURE 1.4
% load chart.

Takt Time

The takt time for an assembly line is the time in which finished units need to be completed in order to meet the customer's demand. Takt is the guideline for how frequently units must come off the assembly line. Takt is calculated as follows:

$$\text{Takt time} = \frac{\textbf{Work time available for a given period}}{\textbf{Customer demand for the given period (number of units)}}$$

Takt time example

Available time = 8 hour per shift – 15 minute break – 15 minute break = 7.5 hrs

Customer Demand = 100 units

$$\frac{7.5 \text{ hrs}}{100 \text{ units}} = \frac{(7.5 \text{ hrs})(60 \text{ mins/hr})}{100 \text{ units}}$$

$$= \frac{450 \text{ minutes}}{100 \text{ units}}$$

Takt = 4.5 minutes/unit

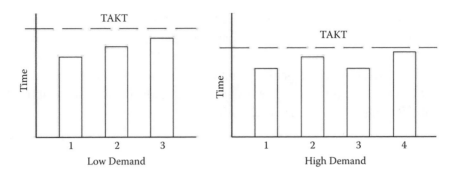

FIGURE 1.5
Level loading at two different production rates (takt times).

work at different stations along the assembly line. Because each operator is responsible for a relatively small amount of work content, the overall skill level required to work on an assembly line or cell is much lower than it would be if each operator was required to build a unit from start to finish.

The level of detail, or the work needed to divide tasks to artificially ensure even work content, and the industrial engineering involved in designing a level-loaded assembly line or cell can be time-consuming and expensive—especially for complex assemblies. Making that kind of investment has always been best suited for products that are expected to have long lifetimes and for which changes to work content are minimal.

Less complex assembly lines or cells can be reconfigured to operate at different rates (takt times) during the life cycle of a unit. If an assembly line, for example, was set up to run with three operators, a second line configuration must be designed that uses four operators when the demand for the product is higher (see Figure 1.5).

Different configurations of an assembly line or cell require different divisions of labor and work content, workstation setup, and documentation (often called *playbooks*). Again, the detail and investment for additional assembly line configurations can be high and cost-prohibitive. Because of this, typically only one configuration is used and operated during the available time to meet customer demand.

When a new assembly line or cell is designed today, level-loading the work content is usually done without hesitation and without considering the age of the method or its original purpose: balancing a paced auto line with operators on both sides.

Self-balancing is an alternative method for flowing products through an assembly line or cell (or any manufacturing or service process). It is a method not born out of an automotive assembly line but a completely different solution to the goal of achieving continuous flow. To illustrate the efficacy of self-balancing, let us first look in more detail at some of the shortcomings of traditional level loading.

2

Shortcomings of Level Loading

The traditional approach to line balancing is *level loading*. This consists of breaking down the work elements needed to make or assemble a product on a line or cell and then distributing those elements as evenly as possible or as is practical among workstations and operators (refer back to Figure 1.4 in Chapter 1). A work element is the smallest distinct step of work needed to make a product and is based on *standard work time* (see sidebar).

ARTIFICIALLY BALANCED

In theory, a *balanced* line is one in which the operators never wait to work on a unit or are required to self-pace (i.e., slow their work so they are not idle or rush to get a unit to the next operator); operators only add value and progress the product through all the steps and workstations. To truly balance a line using level loading, the standard work time for a work element cannot vary, and overall work content must be evenly distributed among the workstations and operators. As such, a truly level-loaded balanced line exists only on paper. It is artificial, and never precisely maintains the designed balance.

Standard work consists of three basic elements: standard work time, standard work sequence, and standard inventory:

- Standard work time: The time in which a trained operator can consistently perform a complete cycle of the work content.
- Standard work sequence: The precise order in which an operator performs all of the work elements.
- Standard inventory: Exact or designated quantity of inventory to perform the work content.

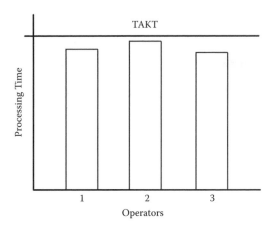

FIGURE 2.1
Level loading, best-case scenario.

Figure 2.1 represents a best-case scenario for level loading, one where cumulative standard work times for each workstation are *approximately* equal.

Note that the work content for each station is not precisely balanced: work is not divided equally among the operators because the work elements cannot be "sliced and diced" to achieve the perfect balance. No two work elements are ever exactly the same, and elements cannot be grouped in a way to make work content at each station exactly the same. And if a product is new to the assembly line or has a short run (is produced infrequently), the variation can be even greater as operators get used to what they are doing. Many factors further negatively affect balance and productivity. Let us take a look at some of the most common factors.

WIP Buffers

Because assembly lines and cells are inherently not balanced, manufacturers need to compensate for the imbalance. Otherwise, all operators except for one, the bottleneck—the slowest operator or station on the line—will be waiting each takt time. Figure 2.2 shows a theoretically perfectly balanced, level-loaded, continuously flowing assembly line in manufacturing.

While this may look like the ideal way to run an assembly line, unfortunately, it is possible only in theory. What more commonly occurs is that engineers design in work-in-process (WIP) buffers of inventory. These buffers are truly that—buffers against the inherent imbalances of the line. So what is wrong with inserting a small amount of WIP to maintain better balance of the operators?

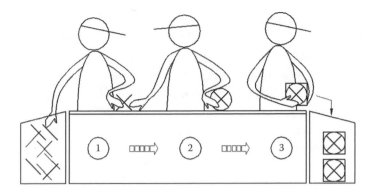

FIGURE 2.2
Theoretically "perfectly balanced" assembly line.

- WIP is inventory, and excess inventory is waste.
- With excess WIP, additional material handling is necessary. Setting down a part just to pick it up again to do the next operation creates more waste (movement). Depending on the type of product, excessive handling also can potentially downgrade or damage the part.
- WIP buffers take up valuable bench or floor space.
- Similar to any inventory, WIP has to be "managed" (counted, sized, safely stored, etc.).
- The WIP buffer contributes to an increase in the production lead time. These small, in-line batches of material slow down the line, and they also can lower the yield if a downstream operation detects a defect (i.e., any WIP sitting in a queue is potentially scrap material).

Look at WIP buffers differently. See them for what they are: waste.

Slowest Operator/Station Sets the Pace

With level loading, the slowest operator or station in your line or cell is a potential *bottleneck*. If your takt time for the line or cell is close to the cycle time of your bottleneck, you must optimize the performance of your bottleneck. The output of the bottleneck determines the output of the line or cell.

In level loading, the slowest operator or station will be the *pacesetter* for the line or cell, keeping workstations in sync, material flowing at a consistent pace, and preventing overproduction. But there are two problems with this: the performance of your line becomes intricately linked to the performance of the bottleneck, and all of the nonbottleneck operators in your line or cell will have excess capacity.

Running a line or cell at a takt time near the cycle time of your bottleneck is unsustainable. Since the bottleneck dictates the output of your line, any unplanned downtime, quality problems, and so on will be difficult to overcome and likely result in late shipments, lost revenue, and an unhappy customer. You cannot constantly risk disappointing your customer. Eventually, you will need to open up more capacity or take fewer orders.

In a line or cell, there is always an operator or station that takes the longest amount of time, statistically or by design. It may not always be the same operator or station. If the cell is running at a takt time near the cycle time of your bottleneck and the cell needs to go a little faster, be a little more productive, and produce at a slightly shorter takt time, it cannot. The cell is always forced to march at the rate of the slowest operator or station, leaving other operators or stations underutilized.

Some lean manufacturing practitioners refer to an eighth waste, that of employees not working to their potential. This waste is often hidden, as operators who do not want to look idle will self-pace in order to work in sync with the bottleneck. Usually, all of the operators in a cell, except the one bottleneck, could contribute more productivity to the cell. My experiences with level loading have shown that the hidden waste of the suppressed operators' potential, combined with the material handling waste of WIP buffers, can be 30% or more of productivity.

Variations in Yield or Cycle Time

Even under ideal conditions in a "balanced," level-loaded cell—one based on standard work in which operators follow the same sequence of assembly steps and take the same amount of time to complete those steps—variations (higher or lower) in yield and cycle time will inevitably throw the cell out of balance and decrease its productivity.

These variations are extremely likely, can be temporary or long term, and come from many sources:

- Slower, new operator
- Faster, experienced operator
- Operator who is struggling on a given day
- Quality problems requiring rework
- Additional work content due to product changes
- Material variation due to product changes or supplier changes

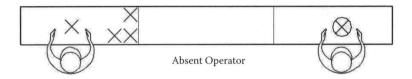

Absent Operator

FIGURE 2.3
Cell with an absent operator.

Since a level-loaded cell is dependent on the even balance of work content among the operators and, as mentioned earlier, is artificially balanced at best, any additional variation will only throw the cell further out of balance.

Absent Operators

Level-loaded lines are designed to operate with a set number of operators, based on the takt time around which the distribution of work was calculated. With an even number of operators, the cell could run with half the number of operators; this would require each operator to be cross-trained in two consecutive operations.

If the cell is designed to run with an odd number of operators or if there is not sufficient cross-training to cover all of the stations, the cell will only run efficiently with the designed number of operators or it will need to be reconfigured. Reconfiguring the cell may seem like a trivial or easy task, given the simplicity of the cells in the figures presented here, but most plants will consist of numerous cells producing complex and varying products, and so a redesign consumes valuable time and resources.

In a cell with an odd number of operators, the absence of an operator with no replacement for that position will cripple the cell (see Figure 2.3).

It will not be able to run as designed (if at all). Consider your operations and what would occur if a cell did not run. If the remaining operators worked on their workstation assignments without the absent operator, it could result in overproduction and all of the waste associated with that:

- Excess inventory: Accumulating inventory in front of the absent operator's station.
- Material handling: Accumulated excess inventory may need to be handled more than once.
- Transport: Excess inventory may need to be moved away from the cell and then back.

- Scrap: Excess inventory may have a defect that is not detected until the absent operator's step.
- Overprocessing: Additional sorting and reading of paperwork.

If the cell could run efficiently with a temporarily reduced staff, it still might be able to deliver a partial order to the customer. But level-loaded cells are rigid in their design and are, therefore, usually run only when fully staffed.

Inflexible, Difficult-to-Adjust Output

A level-loaded cell is usually a nonscalable unit of capacity. It is designed to run at one (or maybe two) takt times. If more (or less) output is needed from the cell, you must adjust the available time and run it longer (or shorter) than a standard shift.

What a level-loaded cell cannot do is easily adjust to match the output needed. Demand is rarely consistent for long, and there are several conditions in which adjustments to output are desirable:

- At start-up or ramping up
- When ramping down or end of product life
- During peak demand, such as seasonal demand
- When overtime is not practical
- At a new takt time
- When demand cannot be leveled or smoothed

Companies often will not design a level-loaded, balanced cell with playbooks to run at more than one takt time due to the difficulties and resources needed to do so. This inflexibility of capacity is a mismatch to inevitable variations in true customer demand.

Silo Mentality

In a level-loaded cell, each individual operator can have his or her unique standard amount of work content, stations, tools, and WIP. The first operator's role is different from the second's, the second's is different from the third's, and so on. If the operators perform their roles while sitting rather than standing, this creates a greater sense of territory—a silo mentality, where the operators are focused on only their work content. They become

FIGURE 2.4
Factory condo.

isolated from the team and less aware of how they contribute to the entire value stream. They might even decorate their work area with family photos and other personal items to create their own *factory condo* (see Figure 2.4). While this might make them enjoy their time at work a little more, it can have the effect of further isolating team members and create barriers to flow.

Repetitive and Monotonous Work for the Line Workers

By definition, level-loaded work is repetitive and can be monotonous. These conditions have remained essentially unchanged since the first assembly lines, with workers performing the same tasks over and over, with little to no variation. As you can imagine, this can result in boredom and employee dissatisfaction, where employees do not feel like their time is honored and their work is meaningful over time. Eventually, these types of work conditions can and have had a negative social impact.

Moving Beyond Level Loading

Clearly, level loading has its merits, as it is used by world-class companies around the globe. It puts in place standard work (a fundamental require-ment of sound operations and one that is still lacking in many firms), highlights waste when deviating from standard work time, and establishes a paced process linked to takt time. In contrast to traditional concepts of building large batches of inventory not sequenced to actual demand, level loading has benefited many companies.

However, in the effort to achieve flow, in particular *continuous* flow, level loading has not gotten us there. Although level loading is *one* solution for line balancing, it is far from perfect. With the hidden and designed-in waiting, inflexibility, and cumbersome process of even setting up a level-loaded cell, it certainly leaves room for improvement. And, yet, there has been essentially no significant improvement to line balancing since its inception.

Could there be a solution to achieving continuous flow that is not just more, better, different than past approaches—but truly is a breakthrough? Let us take a look.

3

A New Paradigm

Balancing lines—or trying to—dates back to Henry Ford and his assembly line in Highland Park, Michigan, nearly a century ago. Because Ford's line was paced, or a conveyor line, he had to level load: in other words, each line operator had to have a similar volume of work (time). Otherwise, operators would either be hurrying to keep up with vehicles moving down the conveyor or waiting idly to perform their operation. We have been level-loading assembly lines ever since in our attempt to achieve flow.

What if Henry Ford had not made cars but some other product that was not assembled on a paced line? Today, we might have a different standard method for line balancing, similar to Ford's line-balancing method but fundamentally unrecognizable. And what if balancing was developed to solve a *range* of problems, not just the problems that occur with a paced line? What if line balancing was developed, first and foremost, to create continuous flow?

CONDITIONS FOR CONTINUOUS FLOW

In lean manufacturing, one of our goals is to achieve flow: the flow of material and information. We also talk about perfection, or the perfect process. In both cases, we are really talking about *continuous flow.*

Continuous flow in manufacturing means that a part or a product moves from one value-adding step to the next. There is no waiting and no additional spacing; each value-adding step moves seamlessly to another. Each step is performed without any self-pacing to try to stay synchronized with other steps. A part is never set down in between steps or placed in a queue, waiting to be worked on. Most of us have never witnessed continuous flow in manufacturing and have only seen a theoretical model (see Figure 3.1).

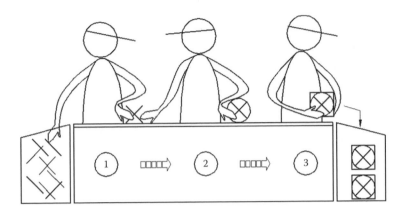

FIGURE 3.1
Continuous flow in manufacturing.

SELF-BALANCING

In its simplest form, self-balancing too is an old concept. Its origins are probably thousands of years old, beginning with our ancestors' need to transport water or move rocks and building materials. To begin to understand self-balancing, picture yourself in the middle of a self-balancing line (see Figures 3.2A through 3.2E).

You are working in the middle of the line, assembling a unit, and progressing down the line until a person downstream from you comes and pulls that unit from you (see Figure 3.2B). That operator takes over the unit you were working on at the station where you were working (without waiting)—and then you walk upstream to do the same for the operator next to you (see Figure 3.2C). These movements are all triggered by when the person at the end of the line finishes a unit and needs a new unit to work on and complete (see Figure 3.2A). Work is pulled from operator to operator up the line, until the operator farthest upstream walks to the beginning of the line to start another unit (when her work is pulled, she has no one from which to pull work) (see Figure 3.2D).

Because the trigger at the end of the line and then subsequent pull of work that ripples up through the line can occur at any time, there is no volume quota or timed objective for an operator—everyone works at their own pace (see Figure 3.2E). The trigger that happens when a unit is completed, combined with each operator working at his or her own pace, creates continuous flow of the unit *independent* of the varying speeds of the operators or complexity of the process steps.

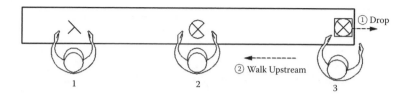

FIGURE 3.2A
End-of-line pull triggered by completion of a unit.

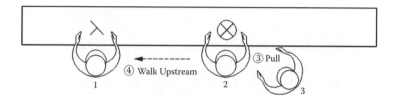

FIGURE 3.2B
First pull taking place.

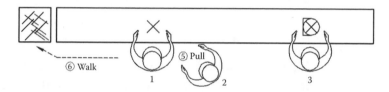

FIGURE 3.2C
Second pull taking place.

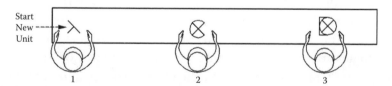

FIGURE 3.2D
Starting new unit at beginning of line.

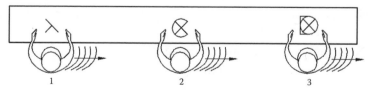

FIGURE 3.2E
Workers work at own pace.

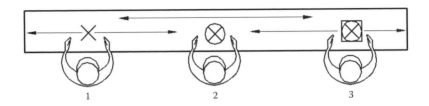

FIGURE 3.3A
Workers in their positions showing range of movement in the line.

	A	B	C	D	E	F	G
1	×	×	×				
2		×	×	×	×	×	
3					×	×	×

Steps (column header group); **Operators** (row header group)

FIGURE 3.3B
Cross-training matrix showing overlap of skills.

When establishing a self-balancing line or cell, operators *are not* assigned to specific workstations for which work content and the time to complete the work are precisely defined. Rather, they are assigned positions on the line or in the cell relative to each other. In Figure 3.3A, the operator at the beginning of the cell is position 1, the next downstream operator is position 2, and so on. In a self-balancing line, operators need to be cross-trained so that they can cover a higher percentage of the total work content than would be required of them on a level-loaded line. For example, if there are three positions in a line as shown, in order to cover their position, each operator has to be trained in *more than one-third* of the line's work content.

Self-balancing lines or cells are exactly that, constantly balanced, always optimizing the division of labor and offering several additional benefits:

- No in-process queues and minimal WIP
- Reduction in material handling
- Communication among the line workers
- Standard work sequence is always followed
- Productivity
- Waiting is eliminated

- Simplicity
- Increased yield

Rules to Standardize Self-Balancing

At first glance, the concept of self-balancing may seem a bit chaotic, with operators moving upstream through a line or cell, grabbing work at any moment from a colleague, and sending that colleague upstream to do the same. But there are basic rules for standardizing the self-balancing process, creating value-adding order and discipline.

Rule 1: Keep Building Progressively Until Someone Pulls from You

When an operator works progressively, he moves downstream to the next station or step in the order it takes to assemble the unit and complete the process. When building progressively, an operator is not concerned with how far down the line he moves or how much work he completes. Only when the downstream operator comes upstream to pull his unit does he stop: the downstream operator takes over the unit, tools, and station without waiting. There is no "Wait until I'm finished." There is only one "finish," and that is by the operator at the end of the line. All other hand-offs, or pulls, are somewhere during the process of building the unit and should take place without waiting (see Figure 3.4).

For nearly all types of manufacturing that take place in a self-balancing line or cell, the pull of work or handoff should not take more than a few seconds. The handoff usually consists of

- Handing over the tool
- Communicating what step you are on
- Moving out of the station and walking upstream

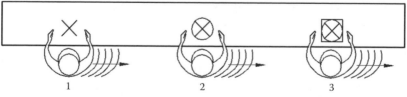

Building Progressively Until
Someone Pulls from You

FIGURE 3.4
Building progressively until the next pull.

For example, let us look at a hand-soldering operation. If an operator is in the middle of soldering when the downstream operator comes up to pull the unit, the handoff would look similar to this:

1. Downstream operator communicates that she is ready to pull.
2. Soldering operator completes the solder joint he is working on (about one second).
3. Soldering operator communicates which soldering joint is to be completed next, per standard work.
4. Soldering operator carefully hands over the soldering iron to the downstream operator.
5. Downstream operator begins soldering while the former soldering operator walks upstream to pull.

Rule 2: When the Downstream Operator Pulls from You, Walk Upstream to the Next Operator and Pull from Them

The operators always know where to go to work on their next unit. No decision needs to be made, and no supervisor involvement is necessary to determine what to work on next. Self-balancing lines are inherently self-directed (see Figure 3.5).

Rule 3: If You Catch Up to Someone When Moving Downstream, Wait

Catching up to someone in a self-balancing line is like tailgating someone in a car: you will need to brake. Waiting is necessary when an upstream operator has completed a station of work and is ready to move into the next downstream station to continue building that unit. But when arriving at the station, she finds that the downstream operator has not completed his work at that station. There is no place for the upstream operator to work (the station is occupied), and so she should wait.

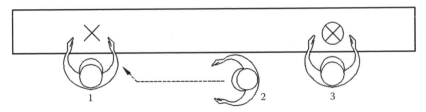

FIGURE 3.5
Walking upstream to pull.

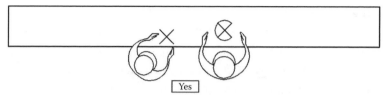

Wait Until Station Is Available

FIGURE 3.6A
"Tailgating," or waiting, to move into the next station.

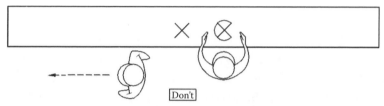

Fumble Unit and Walk Upstream

FIGURE 3.6B
Do not fumble your unit and walk upstream.

Waiting time in a self-balancing line should be short and rare (see Figure 3.6a).

If waits *are not* short and rare, that is a signal that the operation may be a bottleneck, in other words, that the cycle time for that particular workstation is too close to the overall takt time for the line. In that case, consider breaking that station into two shorter, progressive steps to eliminate the bottleneck. In some cases that may not be feasible, and that particular takt time (or shorter takt time) may result in a bottleneck at that station.

Note: Any waiting due to a bottleneck will decrease the productivity of the line, but the line can run this way in order to achieve maximum output of the line by not *starving* the bottleneck of parts to process.

Just as when you are driving in traffic, waiting is not only necessary, but it is still usually the fastest way to get where you are going. The upstream operator should not set the unit down, or *fumble* it, and walk upstream to pull a unit without having been triggered by the downstream operator. People naturally find it difficult to wait for anything, and so they think, "If I set this unit down and go upstream and pull, it will be more efficient." That might be true if the concern was to maximize one individual operator's efficiency, but in lean manufacturing, we try to maximize the efficiency of the entire value stream (see Figure 3.6B).

Setting a unit down or abandoning it causes an extra material-handling step, which will then require some means of identifying the unit, communicating the status of work completed to it, and force someone to come for the unit. It is confusing, wasteful, and contributes both to overprocessing and to inventory pileup. For example, if the downstream operator is temporarily struggling, the upstream operator, having already abandoned one unit, may come back a second time with a second unit to set down. Now there is a small accumulation of inventory in front of the downstream operator, leading to more waste.

Of course, waiting also is waste, but in this case, it is the preferred solution. Why? Waiting with the unit provides visual information that there is a momentary stop to the flow of material. And, if the upstream operator stops, it provides the opportunity for the two operators to communicate that there is a problem. This kind of collaboration is more likely to occur in a self-balancing line because the operators do not have designated stations and tools. If the downstream operator is having difficulty with a unit at a station, the upstream operator also may have difficulty when her work extends into that station. The operators can and should explore why that position in the cell is causing trouble. Communication between operators is encouraged, because it can improve both the process and teamwork in general. This occasional dialogue between operators can help ensure that standard work is being followed. For newer operators, it can be useful to check in with them during this time to see whether there are any problems.

Another reason to wait is because the downstream operator has been at that station for at least as long as the cycle time of the upstream station and has probably almost completed his work. Because of this, the wait is likely to be short compared to takt time. The overall wait time that occurs on a self-balancing line, if any, is just a fraction of the total cycle time to build the unit and can be accounted for when setting up the line around takt time. This slight amount of waiting may stand out as obvious waste in the line or cell. However, it is likely to be the only waste, and the high overall efficiency of self-balancing more than makes up for the slight amount of waiting.

Rule 4: Don't Leapfrog; Everyone Stays in Their Position

As I have already noted, the operators in a self-balancing line are not assigned to specific workstations, but assigned positions relative to each

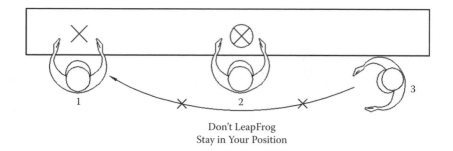

Don't LeapFrog
Stay in Your Position

FIGURE 3.7
Always pull from the next upstream operator.

other. Staying in these positions simplifies the work for the operators. Leapfrogging an operator or randomly changing positions because operators are temporarily "bunched up" would be a nonstandard way to work and would add confusion (see Figure 3.7).

There is a work sequence in which the operators work in a progressive order, and every time an operator reaches the end of the cell, they move to the beginning of the cell and "loop back." This "rabbit chase," as it is often called, also should be avoided. The fundamental flaw in the rabbit-chase work sequence is that operators eventually will start "tailgating" the slowest operator and have to work at the slowest operator's pace—a line or cell paced by the slowest operator will force other operators to work at less than their optimum pace.

Rule 5: Communicate, Communicate, Communicate!

In a self-balanced line, the operators are working as a team—quite possibly for the first time. As with any team, they will need to communicate. When an operator is pulling, for example, he will first need to communicate that he has arrived and is ready to pull. It is a simple communication, usually completed merely by saying, "Pull."

The next communication comes from the operator from whom work is being pulled. For self-balancing to succeed, this operator must communicate critical information, identifying for the downstream operator the last completed task in the standard work sequence and the next task that the pulling operator needs to accomplish (see Figure 3.8).

With some operations, the next task is obvious and the communication can be minimal or even nonexistent. If the pull or handoff occurs during

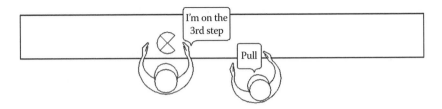

FIGURE 3.8
Communication during a pull.

a complex part of the standard work sequence (see sidebar), the communication can and should be more detailed.

Note: If the takt time for the line or cell is a minute or less, any communication during a pull or handoff procedure must be brief and efficient. The pull or handoff time must be less than 5% of the takt time or else it may result in lost productivity.

Other types of communication may be necessary if an operator needs to leave the line for any reason. An operator may need a short break, in which case she could call in a working supervisor to fill her position. If there are batch or off-line operations near the line (called *backbench*) that require loading or unloading, the operator must communicate that she is temporarily leaving the line. (For more information on backbench, see Chapter 4.)

Note: Any planned time away from the line takes away from the total available time and must be considered when calculating takt time for the line.

DEALING WITH FUMBLES

Self-balancing helps manufacturers strive for continuous flow or the perfect process. But it, too, is not perfect, and you can expect "fumbles" to happen now and again. But you can minimize the impact of fumbles when you understand how to deal with them. Again, a fumble happens when an operator sets down a unit before completion, thereby stopping continuous flow. Fumbles can occur because of both standard conditions and nonstandard conditions.

Adherence to the Standard Work Sequence

No variation to the standard work sequence can exist for long in a self-balancing line. In most other assembly methods, it is often difficult to establish and then ensure that the standard work sequence is being followed. With self-balancing, the line will not work unless the standard work sequence is followed, and the operators are incentivized to adhere to it. Any variation could be revealed during a pull, and would immediately be identified, cause confusion, and stop the line. The pull among operators creates an inherent set of checks and balances to constantly audit that the standard work sequence to build the entire unit is being followed. There is no motivation for the operator to "be creative" or deviate from the standard work sequence. This is a major benefit to self-balancing, because adherence to the standard work sequence is permanently, once and for all, solved for these lines, while traditional manufacturing methods often constantly struggle, operator to operator, shift to shift, to ensure that the standard work sequence is being followed.

STANDARD CONDITIONS FOR A FUMBLE

The standard conditions for a fumble include operators stopping their work for breaks, lunch, and shift changes (see Figure 3.9). At these times, a fumble (stopping continuous flow) is desirable. Because there are no designated in-process queues or WIP buffers in the line to stop processing a unit before completion, by definition the operator will be setting down the unit midstream in the overall process of the build. Just as a handoff can occur at almost any point in the build, so can a fumble.

In lines that are not self-balanced, it is normal for operators heading to a break or lunch to pace themselves and stop their work at a predetermined stopping point: a WIP buffer, in-process queues, or inventory location. If the takt time to which they are working is significantly greater than 1 minute, this method of self-pacing is a hidden waste that can greatly reduce productivity over time.

With self-balancing, there is no need or advantage to self-pace before a break, lunch, or shift change. Since there is "nowhere to get to," an operator

FIGURE 3.9
Fumbles in a line during break, lunch, or shift changes.

can continue to work right up until stopping time: the operator stops, sets the unit down, and upon her return picks up the unit where she left off.

Nonstandard Conditions for a Fumble

Though standard fumble conditions are necessary in most lines, nonstandard fumble conditions are to be avoided. When a line or cell is first created, you are more likely to encounter fumbles, for example, when an operator does not know how to perform a downstream step because of a lack of cross-training for operators in the cell. This problem is most prevalent with products that have only occasional or brief production runs.

To avoid nonstandard fumbles, operators should be cross-trained in a contiguous set of work elements (see Figure 3.10).

If they complete their work elements before the downstream operator pulls them, they will be ready to progress to the next downstream step or station. But in a line or cell that is running a new unit, operators may not have received sufficient cross-training to continue downstream. In other words, they may be trained to perform one or two steps well but not qualified to work any farther downstream.

If an operator gets to a step in which he or she is not trained, one option is to fumble the part, mark the progress with a "Fumble Indicator" (see sidebar), and walk upstream to pull the next part. The Fumble Indicator provides a visual cue for the downstream operator to pick up the fumble during the next pull and within the takt time.

Note: If the operator arrives back at her previous fumble point with a second unit before the downstream operator has picked up the previous fumble, the upstream operator *should not* fumble a second unit. If the fumble is *not*

Steps

Operators		A	B	C	D	E	F	G
	1	X	X					
	2			X	X	X		
	3						X	X

FIGURE 3.10
Cross-training matrix.

The Fumble Indicator

When a fumble occurs, there must be a clear and efficient way to identify and label the fumble. A Fumble Indicator can help ensure quality, standard work, and efficiency. The Fumble Indicator should:

- *List in sequence the standard-work major steps:* A list of standard-work major steps is displayed as a visual aid at each workstation. It consists of a numbered list of concise phrases describing the sequence of assembly for a *trained operator.* It does not contain every precise step at the station, and it is not a stand-alone work instruction. The numbered phrases do, though, represent those discrete steps in the work where a fumble is most likely to occur. Descriptions are brief (phrases rather than sentences), so they can be quickly read, reminding a trained operator of the next step.
- *Include a pointer that indicates which step is to be completed next:* The pointer can be a clip with an arrow or some other robust means of attachment. The pointer is used to identify a discrete step on the standard work list, or it can point to the unit and the place where work is to occur. The pointer should be secure, easy to handle, and not contaminate or downgrade the work area in any way.
- *Be highly visible:* The Fumble Indicator should visually stand out, which can be as simple as using fluorescent colored paper. If the Fumble Indicator also can be displayed upright, it will be easier to see from a distance. The fumbled unit, which is a "stop" to continuous flow, must be easy to identify.
- *Be easy to handle:* The Fumble Indicator must also be easy to place with a unit, remove from a unit, and return to storage.

picked up, this indicates that there is a problem downstream and that the team on the line should stop all work and solve the downstream problem.

When a line or cell is in a start-up mode or is not working to a strict takt time, a fumble caused by inadequate training can provide an opportunity for on-the-job cross-training. When the upstream operator reaches a point where she cannot do the next step, another option is to shadow the downstream operator, receiving real-time instruction from a process expert in the precise work elements that the operator needs to learn—and on real units. After shadowing a few times, the upstream operator should feel ready to try her hand at the first few work elements under the supervision of the downstream operator. This on-the-job cross-training can occur within the guidelines of a quality-control system/TWI (Training Within Industry) program and quickly expand your level of cross-training to reduce fumbles.

Note: Any in-line training performed by operators must be included in the line or cell's takt time calculations.

There are two key benefits to this type of on-the-job training:

- *No redundant training area is needed.*
  In on-the-job training, the same bench, tools, and fixtures used in production are used for training. This saves space, capital, calibrations, qualifications, and so on. There also is no need to sort out "training parts," which are often rejects.
- *Operator motivation is often higher.*
  The work elements that the operators are learning are exactly those they need to learn in order to eliminate fumbling or waiting. When operators can expand their skills one step or one workstation at a time, learning takes place in increments that are not overwhelming. With each learned work step, the cell is closer to continuous flow.

Another nonstandard reason for fumbling is when an operator leaves the line for an emergency (see Figure 3.11). Ideally, a working supervisor or a substitute will take that operator's place. If a substitute is not available, the operator will be forced to fumble the unit when he leaves the line. The operator should then communicate to team members that he is leaving the line and give an estimate of how long he thinks he will be gone. He will then place a Fumble Indicator next to the fumbled unit (if necessary) and leave the line.

If the operator is gone for more than one takt time, his fumbled unit will most likely have been picked up. If so, the returning operator will move back to his position, walk upstream, and pull.

While an operator is absent from a self-balancing line, the remaining operators continue to work as usual. This is in contrast to a level-loaded line that cannot run with an absent operator. The same rules of work

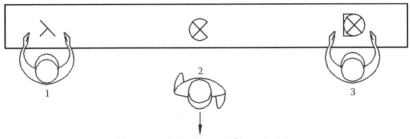

"Emergency" Fumble or Off-Line Activity

FIGURE 3.11
Leaving the line for an "emergency."

progression and pull apply. The remaining operators will temporarily be covering a larger percentage of work in the cell, which could result in cross-training issues (i.e., if an operator has not been prepared to move that far up/down the line). It is important to realize that the cell will run slower (a higher takt time) with the loss of any given operator. If the cell is running to a strict takt time, this lost production will have to be made up through overtime. Provided there is room on the line or in the cell, the lost production also can be made up by adding an extra person to the line (in addition to the returning operator) so that the line can run faster (below takt time) until the cell catches back up to the production goal for that shift.

Yet another nonstandard reason to fumble is to leave the cell to perform some off-line production activities. The same communication to the team should be made when leaving the line. The most common off-line activity is unloading and loading of batch processes (ovens, test stations, etc.). Other off-line activities that cause an operator to fumble include

- Bin replenishment
- Turning off timers or equipment that do not have auto-stop
- Tool repair or replacement

Note: Any planned, off-line activities performed by an operator must be included in the takt time calculations.

Fumble Rules

There are basic rules for dealing with fumbles on the line. The first two rules describe what operators need to do when encountering a fumble, and the third rule describes how to pick up the fumble.

Rule 1: Encountering a Fumble When Moving Upstream

When an operator walks upstream to pull the next unit, and he encounters a fumble between him and the upstream operator, he should pick up the fumble before pulling a unit from the upstream operator. A fumble is a temporary out-of-balance situation. As soon as the fumbled unit is picked up, the line is again in balance (see Figure 3.12).

A fumble will be located in-line at the place where the next work elements are to be performed. If there is a Fumble Indicator (*see sidebar*) with

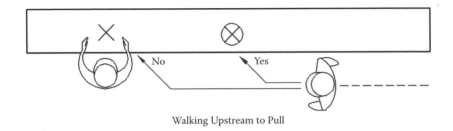

Walking Upstream to Pull

FIGURE 3.12
Walking upstream and picking up a fumble.

the fumble, the operator will read the next work element to be performed and return the Fumble Indicator to where it is stored.

Rule 2: Encountering a Fumble When Moving Downstream

An operator may encounter a fumble when he is working progressively downstream (see Figure 3.13A).

If an operator encounters a fumble under this condition, the operator should fumble the unit he is currently working on and pick up the fumble he has encountered. The operator will fumble at the same step where the previous fumble rests. If a Fumble Indicator is being used, it can then be applied to the new fumble (see Figure 3.13B) (because it describes the same conditions). This rule maintains a first-in, first-out (FIFO) approach to fumbles.

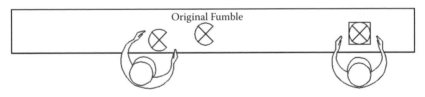

FIGURE 3.13A
Encountering a fumble while progressing downstream.

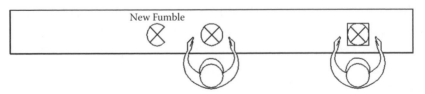

FIGURE 3.13B
Picked-up original fumble.

Rule 3: Returning to a Fumble

This last fumble rule applies to the operator who has fumbled and then returns to the line. If the operator returns to the line within one takt time, it is likely that the unit that he fumbled has not been picked up. If that is the case, the operator returns to the line, picks up his fumbled unit, and continues working (see Figure 3.14A).

If the operator leaves the line for more than one takt time, his fumble most likely will have been picked up. If that is the case, the returning operator positions himself in the area of his fumble between the adjacent upstream and downstream operators, and then immediately moves upstream and pulls from the adjacent upstream operator, putting him back in the line (see Figure 3.14B).

Now that you are familiar with the basic concept of self-balancing, it is time to put those concepts into practice. For simple lines, the concepts can easily be applied. But for more complex assembly lines and processes, the following chapters will walk you through the setup and implementation steps.

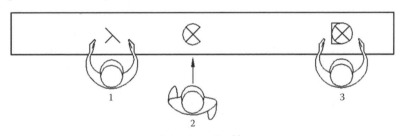

Returning to Fumble
Fumble Not Picked Up

FIGURE 3.14A
Operator returning to a fumble that has not been picked up.

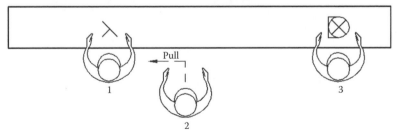

Fumble Picked Up

FIGURE 3.14B
Operator returning to a fumble that has been picked up.

4

Batch and Off-Line Processing

Single-piece flow, or single-part processing, is the ultimate way to manufacture product, because it allows for the least amount of waste in an assembly line or cell. When it is not possible or practical to incorporate single-piece flow for completing a unit or product, batch or off-line processing is required. Batch or off-line processing will add waste to the line, so the goal is to simulate flow with the main line or cell while using the least amount of inventory to run the batch and off-line processes.

Batch, or off-line, processing is primarily needed when a process during the overall assembly operation is over takt time and it is not practical to get the process within takt time: if a unit were to be worked on at the line or cell, one piece at a time, it would be impossible to keep up with customer demand. Many off-line processes also are not safe to incorporate into a line or cell, such as painting and curing ovens, chemical plating, or welding operations.

CURTAIN OPERATIONS

A common method for simulating flow with batch and off-line operations is a *curtain operation*. Curtain operations got their name because flow appears to take place *behind the curtain*, almost similar to a magic show. (There is no need for an actual curtain.)

In a curtain operation (see Figures 4.1A and 4.1B), what occurs "behind the curtain" is the batch or off-line processing. An operator in the line submits one unit for processing *behind the curtain*, and they then reach back through the curtain and pick up what appears to be the same unit after processing. The illusion is that the unit was instantly processed behind the

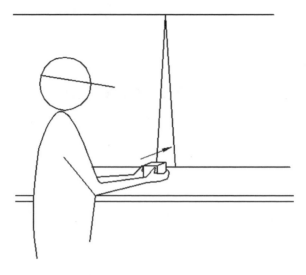

FIGURE 4.1A
Submitting unit "behind the curtain."

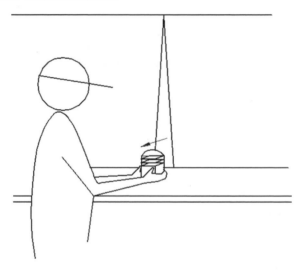

FIGURE 4.1B
Retrieving unit from "behind the curtain."

curtain. Of course, the operator is not picking up the same unit that was just submitted; the unit that was submitted will not be available for pickup until all of the units submitted before it are processed.

Curtain operations are practical methods to keep the operators in the line or cell flowing while batch or off-line processing proceeds in the background. The processing that takes place behind the curtain can take from a few minutes to a few hours. Some curtain operations may even require days of processing and be performed outside the plant.

Curtain operations are performed by personnel dedicated to those processes, not by the operators positioned in the self-balancing line or cell. But for certain types of automated batch or off-line processing, it is possible to have operators from the line or cell load, start, and unload the curtain operation; this station of off-line work would be done without affecting the output of the line.

BACKBENCH

Backbench is a term used to describe batch or off-line processing, similar to curtain operations, except the work is conducted entirely by the operators from the line or cell in concert with the actions of the line or cell. The backbench refers to the bench or workstation that is directly behind the line or cell (see Figure 4.2). When a backbench process is incorporated in the flow of the line or cell, the primary line or cell is called the *frontline*. The close proximity of the backbench work area to the frontline makes it easy for the operators to perform the work.

A straight frontline is the best shape in which to incorporate a backbench. The backbench operation can be lined up directly behind the frontline steps that come before and after the necessary backbench work. In

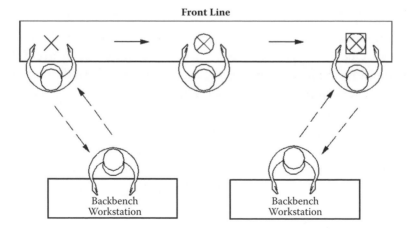

FIGURE 4.2
Assembly line with two backbench processes.

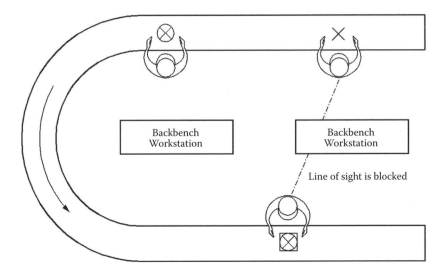

FIGURE 4.3
U-shaped cell with backbench stations.

a U-shaped cell, however, backbench stations might require the design of the cell to be wider than necessary and could disrupt the line-of-site communication across the cell (see Figure 4.3).

IN/OUT TRAY

To keep the frontline operators working efficiently around the batch and off-line processes, you need to minimize the amount of material handling they do. One way to solve that problem is to use an in/out tray to connect the frontline to the backbench process (see Figure 4.4). An in/out tray is a staging area for the WIP inventory both before and after the backbench process. The in/out tray should be placed in line with the cell; if this is not practical, such as for larger parts or batches, the in/out tray may be placed on a separate cart or behind the operator position closest to the backbench station.

Frontline operators place one unit in the "in" tray, pick up one unit from the "out" tray, and continue assembling downstream. The one-in/one-out sequence maintains the standard work-in-process (SWIP) inventory at a constant level. This sounds simple enough, but there are some common mistakes that can cause excess or short SWIP inventory.

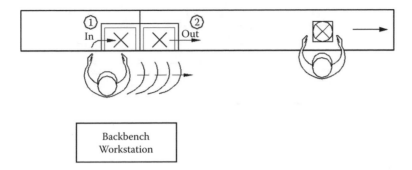

FIGURE 4.4
Backbench process with in/out tray in the line.

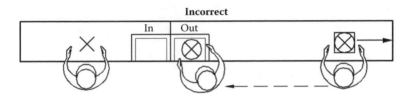

FIGURE 4.5
Walking past SWIP to pull from an operator.

One common SWIP inventory mistake is pulling from SWIP at the wrong time. When an operator is walking upstream to pull from the next upstream operator, the operator must walk past any SWIP inventory. If the operator stops at a SWIP out-tray and pulls from it, the operator will decrease the SWIP level. The only time to pull from a SWIP out-tray is when an operator is building product and progressing downstream, immediately after a unit is placed in the in-tray—remember, "you can only take one out if you just put one in" (see Figure 4.5).

Another common mistake is to add a unit to the in-tray without pulling one from the out-tray (see Figure 4.6). What sometimes happens is that an operator progressing downstream puts a unit in the in-tray and

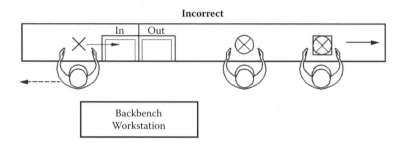

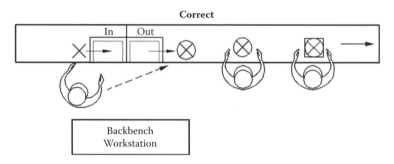

FIGURE 4.6
Progressing downstream past an in/out tray.

then walks upstream to pull from an operator (or start a new unit) instead of progressing downstream, increasing the SWIP level. Remember the one-in/one-out rule and that operators need to build progressively until another operator pulls from them.

The following considerations also will help you incorporate batch or off-line processes without disrupting the flow of the frontline.

- The number of units in the in/out trays should be as few as is practical to keep the SWIP inventory levels low.
- The dimensions of the in/out trays should be sized to hold a single batch of the off-line process.
- If a unit goes outside for processing, the in-tray can be the pack-size quantity that moves to that operation (ideally, it will be the actual packing container, to remove any waste involved in moving parts from the tray to the packing container).
- The in/out trays should not require double or triple handling. The processing tray, cassette, packing container, offloading tray, etc., should always be the first choice for an in/out tray.

GUIDELINES AND RULES FOR BACKBENCH OPERATIONS

Rule 1: Frontline Operators Leave the Line to Perform Backbench Operations on Demand

This is the defining feature of backbench. Unlike a curtain operation, in which frontline operators do not participate, there are no operators dedicated or assigned to the backbench. The work done at the backbench is done *on demand* by a frontline operator who returns to the frontline once the work is done. When units are available for processing at the backbench station and the station is available, a frontline operator begins work on those units at the backbench.

There are different levels of automation for backbench processes. For manual load and unload, frontline operators must be aware when a machine needs to be attended to and moved into that assignment as needed. If the machine has auto-stop or auto-unload capabilities, it is easier for an operator to focus on the frontline operations. For example, an empty out-tray on the frontline could be the signal to transfer parts from the backbench to the frontline.

While a frontline operator is performing backbench work, that operator is technically off the frontline: in other words, when a downstream operator walks upstream to pull, they do not pull from an operator that has moved to the backbench. The downstream operator does not walk into the backbench area but pulls from the next upstream operator in the frontline. The backbench area is not a pull area. Pull only occurs on the frontline.

Rule 2: Operators Communicate When Leaving for the Backbench

Team member communication is critical on a self-balancing line. If an operator must leave the frontline to go to the backbench, she must first communicate to the team:

- When she is leaving
- To which backbench process the operator is going (more than one backbench can support a frontline)
- Approximately how long she expects to be gone

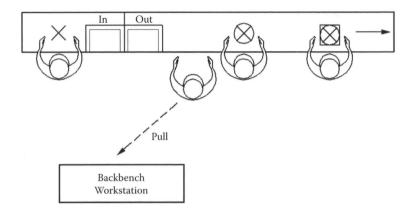

FIGURE 4.7
Operator leaving for backbench processing during a pull.

This communication lets the team know that other operators will not be pulling from the operator who has gone to the backbench. While an operator is away from the frontline, the remaining operators will cover more of the work on the frontline. Operators need sufficient cross-training so that they can cover a wider swath of work for these adjustments to the frontline. But these on-the-fly adjustments also illustrate a key advantage of self-balancing: the line does not require an engineer or other supervision to get rebalanced if an operator leaves. The line will remain in balance.

The unit the operator was working on before he went to the backbench will be fumbled, or left behind, and when he leaves the frontline to go to the backbench, he needs to let the team know its status. The best way to communicate this information is with a Fumble Indicator (see Chapter 3). Although simply telling the team the status of a unit is helpful, it should not replace the use of a Fumble Indicator. Often the next step to be performed to a fumbled unit is obvious, but good communication in a self-balancing line or cell cannot be overdone.

A convenient moment to leave a frontline to go to the backbench is in the middle of a pull (see Figure 4.7). If the operator going to the backbench goes during a pull, the operator will not have a unit in hand, there will be no fumble, and communication about the status of a fumble will not be necessary. But the operator should not wait for a pull from the downstream operator before he proceeds to the backbench if:

- The backbench process needs to be manually stopped.
- Waiting could cause a stock-out for the frontline.

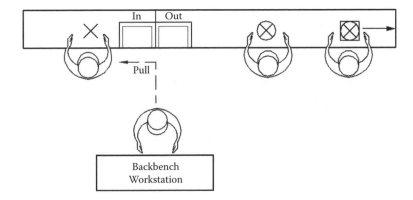

FIGURE 4.8
Operator returning to line position from backbench.

Rule 3: Reenter the Frontline in the Same Position

After the backbench work is complete, the operator returns to the frontline without delay, returning to the same position he was in when he left the frontline. If the operator left a fumble and it has not yet been picked up, the operator will pick up the fumble and continue. If the fumble was picked up (or if there was no fumble) (see Figure 4.8), the operator will return to his position, walk upstream, and pull.

TYPES OF LOADING AND UNLOADING

Backbench processes can address a variety of units and types of work, but they typically are categorized as one of three types:

- Batch
- Single unit
- Multiple station direct loading

Batch is the most common form of loading and unloading of the units for a backbench process. The units are loaded and unloaded in a batch. The amount of SWIP inventory requires one batch being processed, one batch in the in/out tray, and a small amount (if any) of buffer stock. Some examples of batch operations are oven-curing, brazing, burn-in, and testing.

Single-unit loading may be processing in a batch, but units can be added one at a time (i.e., a unit is loaded one at a time, while multiple units

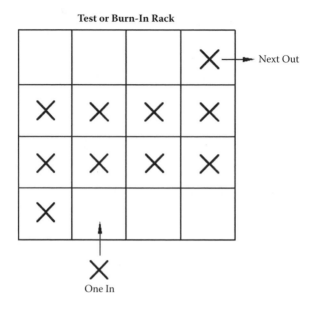

FIGURE 4.9A
First-in, first-out (FIFO) loading and unloading.

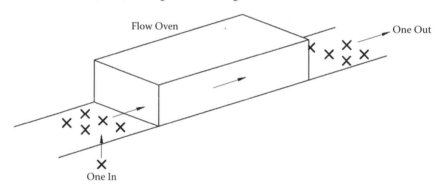

FIGURE 4.9B
One-in, one-out loading and unloading.

in-process are gradually processed in the batch). Single-unit loading has the least amount of SWIP inventory because there is no need to accumulate an entire batch before loading (see Figure 4.9). For example, a flow oven is commonly used for single-unit loading. Convection ovens can be used if opening and closing the oven door does not affect the process. Burn-in testing, where banks of units are tested at the same time, also can be designed for single-unit loading and unloading.

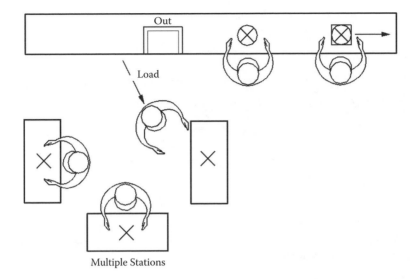

Out

Load

Multiple Stations

FIGURE 4.10
Single-unit direct loading of an off-line station. No in-tray required.

Multiple stations with direct loading are needed when the processing time is above takt time. Multiple stations are moved to the backbench even though processing occurs one unit at a time at each of the backbench stations. Multiple stations can be automated or manual. And rather than use an in-tray, multiple stations can be loaded directly by the frontline operator. Direct loading eliminates a material-handling step and eliminates the need for accumulating SWIP inventory in an in-tray.

If the multiple stations are manual processing, operators performing the work will be positioned at backbench stations. When a unit is finished, the backbench operator brings the unit to the frontline. A small buffer of units can be in an out-tray to keep the frontline flowing. The backbench operators will flex in and out between the multiple stations and the frontline to balance their workload.

Inline batch processing can be done if the units are small and if the placement and use of equipment does not cause excess walking for operators (see Figure 4.11). When operators come to the batch operations, they load one unit, unload one unit, and then continue assembling. But because operators will pass the inline batch process every takt time, any inline batch processing station that requires excessive walking would build in waste and is usually an unsatisfactory design for self-balancing.

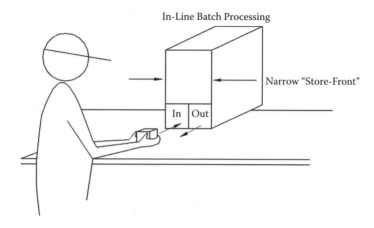

FIGURE 4.11
One-in one-out, and continue.

Batch operations need not disrupt flow or cause excessive waiting. Backbench operations allow the operators to perform the off-line work without the need for additional staff and supervision. When subassemblies feed into the main line, self-balancing can also be used to link and coordinate the two lines. In the next chapter, we will take a look at how to do that.

5

Subassembly Feeder Lines

In an ideal assembly design, every component of a product should be part of a frontline process: the bill of materials (BOM) should be a single-level list of items (flat), and any subassemblies, if necessary, should be built in-line as needed (vertically integrated). Unfortunately, it is not always practical or feasible to put everything in one line or cell. Subassembly feeder lines are sometimes needed to supply parts to a final assembly line.

With traditional balancing techniques, a separate subassembly line is coupled to an assembly line by a buffer of inventory between the two lines, ensuring that there is enough material to keep the main line operating. This, of course, is waste but is necessary because the two lines are not perfectly balanced because of statistical variations between the two.

The same issues that are encountered when attempting to balance an assembly line with level loading are also present when balancing a line with a subassembly line: what usually occurs is that extra inventory buffers are incorporated and additional supervision is required to manage the operators between the lines. Unlike traditional techniques, self-balancing processes can be used to balance a frontline with a subassembly feeder line, without the extra inventory or additional supervision. In a self-balancing line, the frontline and feeder line are arranged so that the subassembly line "feeds" the frontline, linking the two and allowing operators to flex seamlessly between the two lines (see Figure 5.1).

RULES FOR SELF-BALANCING SUBASSEMBLY FEEDER LINES

Here is how to think about the relationship between the frontline and the subassembly feeder line. The frontline is the "river," flowing a unit, part,

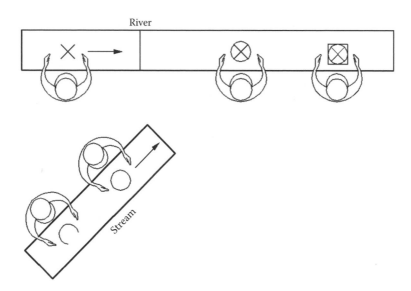

FIGURE 5.1
Layout showing self-balancing frontline and feeder line.

or product to completion. A subassembly feeder line that links to the river is a "stream." Considering this, "upstream" can literally mean moving up the feeder line, and "up the river" or "down the river" means moving up or down on the frontline.

Rule 1: Going Down the River

For an operator going down the river, progressive assembly is the way to move down the line, just as in a self-balancing line without a feeder line. When an operator arrives at the pickup point for the subassembly line (if there is a subassembly available), the operator continues assembling "down the river" (Figure 5.2).

The second scenario when going down the river is if an operator arrives at the pickup point and there is no subassembly available. The operator cannot continue down the river without the subassembly but does not have to wait. Instead, the operator fumbles the unit that she is working on at the frontline and walks upstream to pull and retrieve the subassembly from the subassembly feeder line (see Figure 5.3). After completing the subassembly, the operator with the subassembly in hand will rejoin the river to pick up the fumble and continue assembling. (Note that when

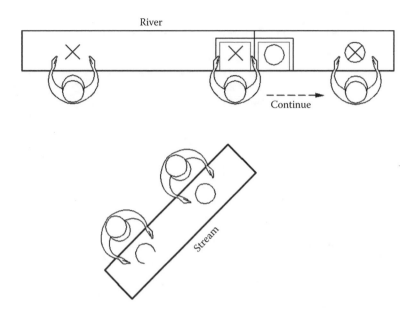

FIGURE 5.2
Progressing down the river when a subassembly is available.

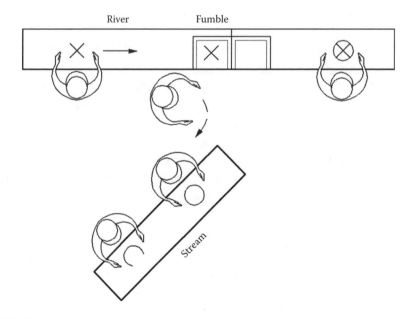

FIGURE 5.3
Progressing down the river when a subassembly is not available and then going upstream to pull.

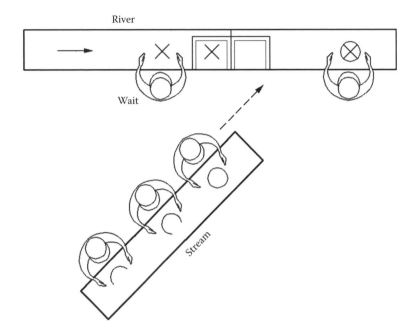

FIGURE 5.4
Waiting for frontline operator to return with subassembly.

fumbling at a pickup point, a Fumble Indicator is not needed because there is no variation in the next step of work.)

The third scenario we see when going down the river is when an operator arrives at the pickup point and there is already a fumbled main unit. The operator arriving at the fumble must wait until the operator that fumbled the unit picks up the fumble (see Figure 5.4). A fumbled main unit indicates that a frontline operator has gone upstream to pull a subassembly. Waiting for that operator to return to the frontline maintains the positions of the frontline operators.

Rule 2: Going Up the River

After each completed unit at the end of the river, each frontline operator will be walking up the river to pull the next unit. For the operator walking past the pickup point, there is another set of rules to manage the flow.

When a frontline operator walks up the river past the pickup point and there are no main units or subassembly fumbles, the operator keeps walking upriver to pull from an operator (see Figure 5.5). The operator has no choice because there are no fumbles to pick up.

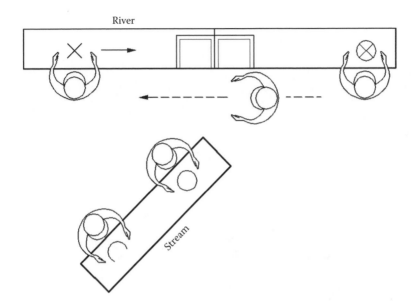

FIGURE 5.5
Walking up river past empty pickup points.

The second scenario when an operator is walking up the river happens when there is a fumbled main unit. The operator cannot pick up the fumbled main unit because there is no subassembly to go with it. The operator must, therefore, go upstream to pull and retrieve a subassembly from the feeder line (see Figure 5.6). When the operator completes the subassembly, the operator can pick up the fumbled main unit and continue down the river.

The third scenario when an operator is walking up the river happens when there is a fumbled subassembly. The operator will walk past the fumbled subassembly and pull from an upriver operator (see Figure 5.7).

RULES FOR SUBASSEMBLY FEEDER LINE OPERATORS

When a subassembly feeder line is linked to the main line through self-balancing, the operators are assigned to a position either on the frontline (river) or subassembly line (stream). As the operators flex back and forth between the two lines, it is important that the operators understand and maintain their positions. There are a few rules for subassembly feeder line operators to manage their positions and to keep the frontline flowing.

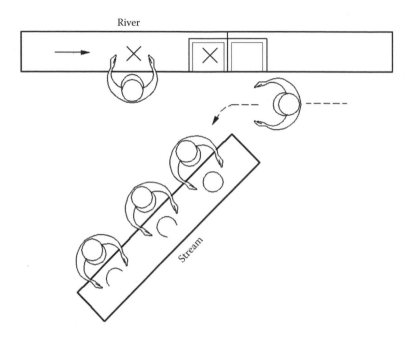

FIGURE 5.6
Walking up river and seeing a fumbled main unit.

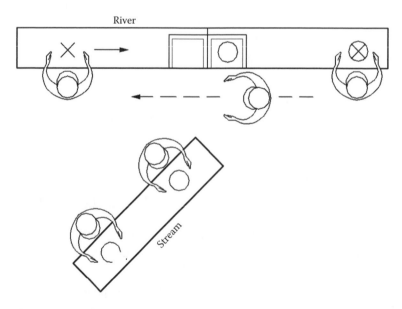

FIGURE 5.7
Walking up river past a fumbled subassembly.

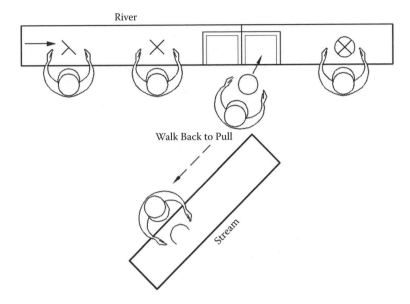

FIGURE 5.8
Subassembly operator delivering subassembly to the line.

Rule 1: When Frontline and Feeder Line Are Synchronized

The ideal self-balancing scenario is when the two lines are synchronized (a subassembly is at the frontline as needed, with no extra subassemblies). The subassembly line will be supplying the frontline just-in-time. If the subassembly line operator delivers a subassembly to the frontline and there are no fumbles at the pickup point, the subassembly operator should fumble the subassembly and return to the subassembly line to pull (see Figure 5.8). The frontline has a subassembly ready to go with the next unit just-in-time.

Rule 2: Encountering a Fumbled Subassembly

The only other possible scenario for a subassembly feeder line worker is delivering a subassembly to the pickup point and finding an already-fumbled subassembly. The operator should then wait for the fumble to get picked up (see Figure 5.9). This scenario is an indication that the subassembly line is ahead of the frontline.

Once the fumbled subassembly is picked up by a frontline operator, the next subassembly should be fumbled to the pickup point. Because the subassembly feeder line has gotten ahead of the frontline, the subassembly operator should now walk up the river and pull (see Figure 5.9B).

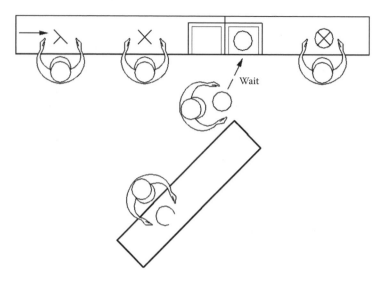

FIGURE 5.9A
Subassembly operator waiting for prior subassembly to get picked up by the frontline.

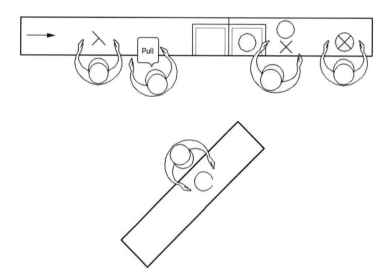

FIGURE 5.9B
Subassembly operator walking up river to pull.

The subassembly operator will return to the subassembly feeder line after the first pull from a frontline operator.

Arranging a subassembly feeder line that self-balances with a frontline can be challenging. The rules and scenarios for the operators can take practice and debugging, but the benefits are worth the effort:

- Productivity is increased because of self-balancing and minimal waiting.
- The buffer of inventory under any circumstance is the absolute minimum—one—while still allowing both lines to flow.
- The two lines manage themselves without additional supervision; the two lines communicate as if they were one line.
- Problems with one line or between the two lines are identified immediately without excessive buildup of inventory.

Now that you are familiar with the self-balancing concept of linking up off-line processes and subassembly feeder lines, let us look at how to debug the lines to create and sustain continuous flow.

6

Debugging the Line

Self-balancing is all about achieving true continuous flow. A line or cell that is not achieving continuous flow must be debugged until the line or cell flows. Debugging begins at the start-up of a line and, similar to the start-up of any complex process or machine, requires observation, patience, and a commitment to improve. It involves the engineers, line supervisor, and the assembly line team.

The initial step for debugging a line or cell is to walk the process and observe one operator assemble one unit from start to finish and return to the beginning of the cell—in other words, observe one complete cycle. This method—walking through and observing the actions of a single operator—is the simplest way to observe what is really occurring and enables you to see whether the line is ready for one-piece flow and self-balancing with multiple operators. (If a single operator is not trained in all of the process steps for this observation, additional operators can be placed in the line—acting as if they are one operator—and they can hand off a unit to one another to complete assembly.)

When watching the one operator, the objective is to observe whether one part flows continuously all the way through the line. Look for the following:

- Progressive assembly: Does the operator move progressively down the line with no backtracking?
- Material handling: Are the parts easy to handle and sufficiently spaced throughout the line?
- Part and tool presentation: Are the parts and tools presented in the correct order and easy to reach?

One operator working in a cell is always balanced; a single operator will not wait or produce too fast. Once one-piece flow is established at this basic level, additional operators can be added to the line to begin self-balancing. As operators are added and pull without waiting, additional debugging may be necessary to address:

- Standard work: Is the sequence of each step defined and clear?
- Bottlenecks: Are operators waiting to move into a station?
- Standard work-in-process (SWIP) inventory: Is SWIP inventory (if necessary) sized correctly? Does the SWIP level remain constant?
- Crowding: Are the operators too close to one another at any part of the line?
- Excessive walking: Do any of the walking distances stand out as long and wasteful?
- Operator return upstream: When the operators walk upstream, are there any tools, fixtures, trays, or carts that need to be returned?
- Handoffs: Are the handoffs occurring without waiting? Is it easy to transfer the unit, tools, and station to the pulling operator?

Debugging activities for a fully staffed line will focus primarily on the application of standard work, dealing with any bottlenecks, and managing SWIP inventory.

STANDARD WORK

For any step on a line or cell, standard work should explain to operators the proper sequence to perform a step. Debugging for insufficient or incorrect standard work at specific work steps can occur, as noted, during the single-operator walkthrough, when we see an operator struggle with a step as they search for the right sequence or parts. But debugging also will occur when the self-balancing line is under way with all operators. Key moments to observe are when operators pull from other operators during the assembly. If there is any confusion or variation in standard work, it is revealed immediately during these handoffs. When a pull occurs, the next step during the handoff must be absolutely clear to both operators. And because handoffs can take place throughout the line, the evidence of standard work will be thoroughly tested.

During the start-up of a line or cell, it is not unusual for a downstream operator to have a problem at a handoff. The downstream operator, who is pulling, may be confused by the next step of work to be performed with the unit, or he may be unsure and need to check whether or not previous steps were completed by the upstream operator. Confusing and repetitive steps often occur because of a lack of a clearly defined standard work for the step (i.e., if the step is not adequately standardized, it is difficult to communicate what needs to be done during a handoff). This is especially true of inspection steps.

Inspection steps (such as examining a unit to check quality) are often intentionally but erroneously not defined using standard work. The belief is that an inspection step is either too difficult to document, or that it is not necessary to document a standard work sequence, and so it should not be interrupted by the pull from a downstream operator—that is, you should simply complete the inspection and then hand off the unit. But standard work can and should be implemented for inspections that can be broken down into precise instructions—identifying each activity, the order of activities, and the duration of activities—which allows operators to quickly follow the sequence and communicate the next activity of the inspection step when a pull occurs. With the right standard work in place, what before may have been a random inspection process is now standard-ized and repeatable. A handoff can take place without redundant checking or confusion.

The more you can standardize inspections, the more likely you will be to integrate inspection steps—including end-of-the-line inspections—into self-balancing processes. Instead of relying on "quality control" to check units at the end of the line (often at random and from finished units that are ready to ship), those inspectors can be incorporated into the self-balanced line and work in coordination with the team as another operator. Your effort to incorporate inspectors may encounter some political battles (inspectors may not want to assemble products, for example), but it reduces waste and improves overall line productivity. Final inspection should be part of self-balancing.

BOTTLENECKS

Bottlenecks are common during start-up of a line or cell or when takt time changes for a line. In a self-balanced line, a bottleneck is a process or step that is near, at, or above takt time. You can easily identify bottlenecks

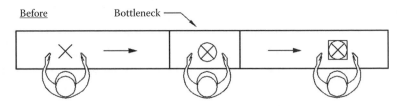

FIGURE 6.1
Waiting upstream from a bottleneck process.

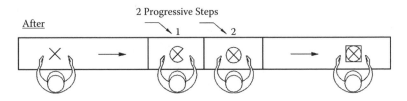

FIGURE 6.2
A bottleneck process after being broken up into two progressive steps.

during debugging when you see an operator who must consistently wait to move into an occupied downstream station (see Figure 6.1).

A bottleneck in the line puts the ability to deliver according to customer demand at risk as downtime accumulates and productivity for the line or cell decreases due to the waiting. One method for debugging a bottleneck is to break up the process into two or more progressive steps (see Figure 6.2).

Instead of occurring at one workstation, the process is distributed over two separate workstations; in other words, turning a step that is at (or near) takt time, to two steps that each average half of takt time. While this makes the line longer and uses more bench/floor space, it eliminates waiting and ensures that each separate step will be below takt time.

By breaking up a step, the unit will progress through the stations more quickly. But note that the material handling of the unit will increase, and this will require careful design of the steps to minimize productivity loss and any downgrading of the unit.

SWIP INVENTORY

SWIP inventory is a standard, fixed-quantity of inventory maintained by one-in/one-out. SWIP is important to achieving continuous flow when a frontline connects to batch or off-line processes. SWIP inventory should be kept to a minimum, but some conditions will necessitate larger SWIP inventories, such as batch processes with subassemblies that need to cool or dry prior to handling (see Chapter 4). Debugging identifies whether or not SWIP inventory is sized incorrectly or if operators are unclear about how to handle it, conditions that impede continuous flow.

As long as one-in/one-out is followed, the SWIP inventory level will not vary. Managing SWIP inventory involves training and repetition to ensure that operators understand and follow the standard work. Debugging for SWIP is most likely to be needed during start-up of a cell, and is triggered by a variation of the SWIP inventory level.

SWIP Increasing

If the SWIP inventory level is increasing, it indicates that an operator put one or more subassemblies in and did not take one out and continue down the line. For some reason, the operator walked back up the line before a downstream operator pulled from the operator.

Fortunately, it is easy to identify and correct excess SWIP inventory levels. The best way to minimize excess SWIP inventory is to mistake-proof the SWIP storage area by not providing extra locations to place a unit. If the SWIP inventory level is four units, the tray only holds four units. An operator would have to put an extra unit somewhere else, which would visually stand out. Any accumulation of excess SWIP inventory easily would be identified and alert the operators that they have made a mistake.

Counting the SWIP is another simple way to identify excess SWIP inventory. The SWIP inventory should be presented in a way that makes it easy to count and control. During start-up and debugging of a cell, frequent cycle-counting checks can identify any variation in SWIP inventory levels.

SWIP Decreasing

A SWIP inventory level below the standard also indicates a breakdown in one-in/one-out, and this could lead to a stock-out that stops the line. A lower-than-standard SWIP typically reveals that an operator walked upstream to pull and, instead of pulling from another operator, stopped and pulled from a SWIP inventory tray, and the SWIP inventory decreased by one. If the same mistake is repeated, a stock-out eventually will occur.

The same methods to identify excess SWIP inventory can be used to identify shortages. But a shortage might not *visually* stand out like excess SWIP inventory. An extra empty spot in an in/out tray is not as obvious as excess units accumulated outside of a tray. Because of this, cycle counting and then immediately making corrections to increase the SWIP level can prevent shortages.

The time and actions needed to correct a shortage or excess of SWIP inventory disrupts flow in the line, but corrections are necessary to immediately restore the inventory levels. To restore these levels, you need to reverse whatever action caused the SWIP level to deviate from the standard. For example, if there is a shortage, operators must add to the SWIP until the inventory level is back to the standard. To correct for excess SWIP, operators must pull from the SWIP—without adding to it—until the inventory level is back to the standard.

This back-and-forth adjusting of the SWIP inventory levels extends the time needed to debug a cell because the operators spend time working in a nonstandard way to correct the problem (i.e., breaking from one-in/one-out). It is important to quickly identify any variation in SWIP levels, stop the line, and correct the variation. But, equally important, operators must learn during debugging what caused the variation to the SWIP inventory and how to prevent reoccurrences. The team needs to review the standard work steps, their practices, and any additional mistake-proofing improvements.

ADDITIONAL DEBUGGING

While debugging should focus primarily on requirements for a self-balancing line, debugging must also address common manufacturing objectives:

- Safety: Are there hazards or risks to which the operators are exposed?
- Ergonomics: Are operators performing repetitive or awkward motions that could result in pain, discomfort, or injury?
- Documentation: Is required documentation accessible and easy to read?
- Transactions: Are all transactions and data gathering simple, timely, and error-proofed (e.g., barcode scanning, not writing)?
- Lighting: Is lighting adequate to perform quality work?

During debugging, any disruptions to continuous flow will stand out as fundamental wastes and must be eliminated. Debugging of a self-balancing line can take from hours to weeks. It is important to get through the debugging process as quickly as possible, so that you keep up with demand and the operators and the rest of the team do not get frustrated and want to revert to "the old way." When possible, an expert can help shorten the debugging process. You know debugging is successful when operators on the line, with the occasional support of a supervisor, keep the line flowing. As with all operations improvements, ongoing observation and identification of wastes—no matter how small—will continuously advance the productivity of the self-balancing line.

7

Horizontal Arrangement of Parts and Tools

Parts in a self-balanced line should be presented or arranged in a horizontal fashion: in the order they are to be assembled and in one contiguous, horizontal line down the line or cell (see Figure 7.1). Because operators are progressing down the line as they assemble, parts, if they fit, must be presented in front of them at their point of use on the line.

If the same part is needed at two or more locations during the assembly, it should be presented at each location where it is needed (not in one spot, from which operators must reach across to grab parts). This means that there may be redundant locations for a specific part. Even if the part is assembled in two locations that are separated by only one step involving a different part, it still should be presented twice (see Figure 7.2).

The tools used to assemble units also need to be presented in a horizontal arrangement. This ensures that the proper tool is available at the location in the line where it is needed, along with the parts for the unit that is being assembled. Again, if a tool is needed at more than one station on the line, a tool should be placed at every station where needed.

Often, when setting up a line, cell, or workbench, a shadow board is used to arrange tools (see Figure 7.3). Shadow boards generally are an excellent technique to keep track of the tools needed in an area (i.e., if the shadow is empty and the tool not in use, then the tool is missing). In a self-balancing line, however, a traditional shadow board containing many tools is not the preferred way to arrange tools: shadow-board tool presentation in a self-balancing line has space for only *one* tool needed for the one station (see Figure 7.4). Each tool needed on the line should have its own shadow board. Another functional way to store tools in a self-balancing line is to attach them to a retractable, overhead line (see Figure 7.5).

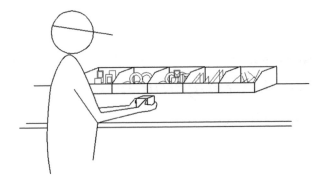

FIGURE 7.1
Horizontal parts presentation.

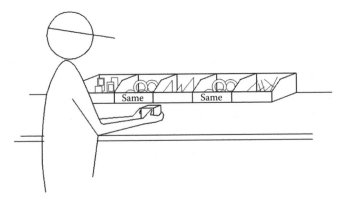

FIGURE 7.2
Same part presented twice.

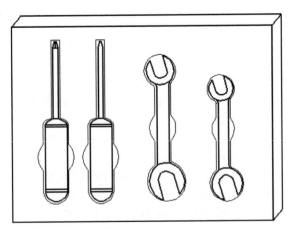

FIGURE 7.3
Traditional shadow board for storing tools.

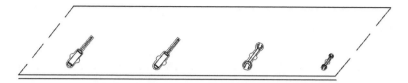

FIGURE 7.4
Tools stored at point-of-use horizontally.

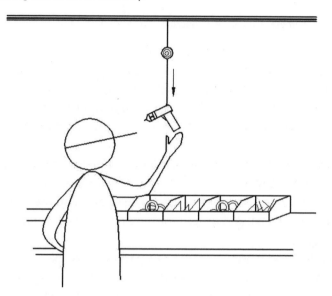

FIGURE 7.5
Retractable tool stored vertically.

BENEFITS AND POTENTIAL DISADVANTAGES OF HORIZONTAL PART AND TOOL PRESENTATION

The horizontal presentation of parts and tools in self-balancing is a great way to help ensure that standard work is followed. Progressing down the line and assembling in the order in which the parts are laid out is a nearly mistake-proof method to ensure the proper sequence of assembling and use of the proper tools. Vertical arrangement of the parts (stacking parts above and below each other), on the other hand, only adds confusion to the sequence of assembly.

Anything that helps reinforce standard work prevents mistakes and improves quality, because an operator is less likely to assemble in the wrong sequence or omit a part or component. This horizontal arrangement

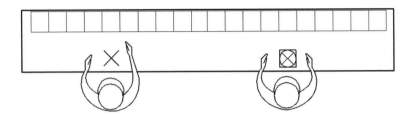

FIGURE 7.6
Status of work progress indicated by horizontal location in the line.

of parts and tools also can assist productivity and safety. For example, there should be no reason for an operator to overreach or extend awkwardly to get to the right part or tool, since it is within arm's reach. And because operators can be quite close to one another while working on slightly different steps in the assembly, it ensures unimpeded movement and saves time.

Another benefit of horizontal part and tool presentation is that it simplifies training. A new operator is trying to learn many things when first starting a new process. If the sequence of assembly and the correct tool to use is obvious, visual, and simple, the operator can be confident and learn more quickly.

The visual presentation of tools and parts also enables an operator or supervisor to immediately assess the status of the overall unit build. The location of the unit and the operator in the cell precisely indicates what parts have been incorporated, the extent to which the unit has been completed, and what is next for each unit in the line (see Figure 7.6). This may seem like a trivial point, but it can be very beneficial to the supervisors managing the cell. In a traditional line or cell, the visual or unit progression often is not always clear, and the exact status of a unit can be hidden and difficult to ascertain because assembly of a unit may stop during the build for extended periods of time.

During breaks and shift changes, a unit will be fumbled and placed on the bench in front of the next part or step in the assembly. When an operator returns to work on the unit, the horizontal parts and tool presentation make the next step in the process clear. The Fumble Indicator (see Chapter 3), which lists the major standard work steps, is another reinforcement to ensure proper assembly.

Many of the benefits of horizontal part and tool presentation also could be realized on traditional lines. But unlike a traditional line, the success of a self-balancing line or cell is dependent on horizontal part and

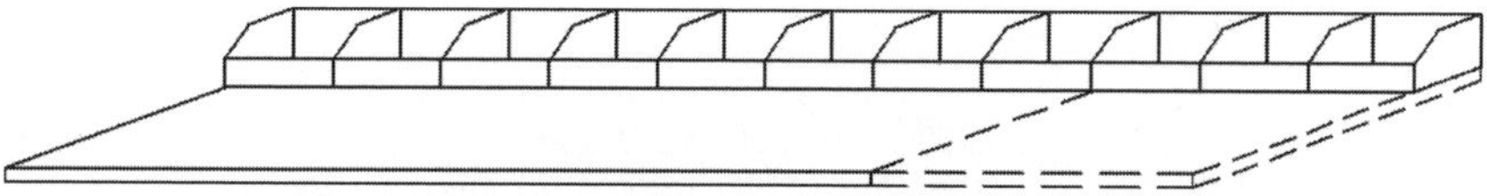

Length of Horizontal Presentation Parts

Length of Line Required for Assembly   Additional Length of Line
                                              to Match Presentation of Parts

FIGURE 7.7
Horizontal presentation of the parts can exceed the length of the line needed for assembly.

tool presentation, and the benefits of this approach are *always* realized. However, despite the many benefits of horizontal part and tool presentation, it is important to understand potential disadvantages.

Managing multiple and redundant locations for storing parts will add some carrying cost and material-handling labor, as well as the cost of purchasing redundant tooling, fixtures, etc. The physical arrangement of the parts and tools also may make for a long line, cell, or workstation, particularly if there is a large parts list (bill of materials); the actual width of the horizontal presentation of parts could exceed the amount of room needed for the assembly of the unit (see Figure 7.7).

To help place horizontal presentation of the parts within a minimum station width and, thus, keep line lengths manageable, bins should be sized as narrowly as is practical to match part width. But, first and foremost, bins should be easy to use, and it should be safe for operators to remove parts from the bins.

ADDITIONAL ROWS OF PARTS FOR MIXED-MODEL LINES

If the parts and the bins that hold them are not too tall, two rows of bins can be presented in a single cell (see Figure 7.8). This is done in a mixed-model cell where two or more products are run down the same cell.

As with most part presentation, gravity-fed racks should be used that allow parts to be pulled from the front and replenished from the rear (see Figure 7.9). This method ensures first-in, first-out for the parts being assembled when multiple bins are being stocked on the line. When a bin is emptied, a return ramp can be used to return the empty bin to behind the line or cell, where part replenishment occurs. A material handler

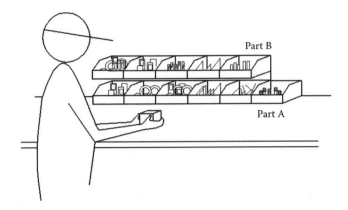

FIGURE 7.8
Parts presented for two different products assembled down the same line.

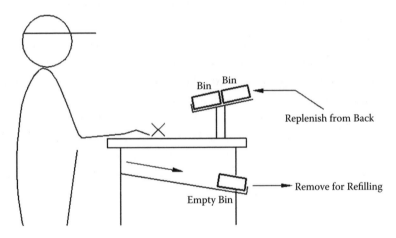

FIGURE 7.9
Gravity fed first-in-first-out bin system.

(sometimes called a "water spider") can then pick up the empty bins while they are making their rounds to replenish parts. This is all done from the rear of the line or cell with no disruption to the operators working in the cell.

A two-bin system is a simple way to stock the parts in the line. The bin quantity is sized such that an empty bin is replenished before the bin from which the operator is working is emptied. In other words, the time it takes for an operator to use the parts in the bin is slightly longer than the time it takes a material handler to replenish the other bin (pick up the empty bin and place a full bin at the cell). No complex math is needed,

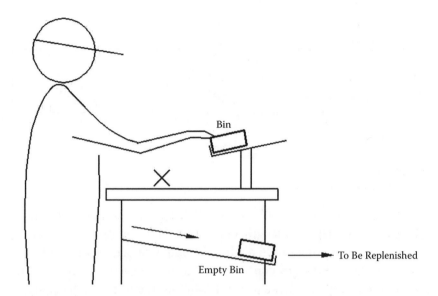

FIGURE 7.10
Two-bin system.

but, obviously, the faster the replenishment times, the smaller the bin quantities (see Figure 7.10).

QUICK PRODUCT CHANGEOVER OF A CELL

If a cell or line is designed and set up with progressive, horizontal assembly, whether it is a self-balancing cell or a traditional line, quick product changeover will be possible. When changing over a cell from one product to the next, the bins and tools will often need to be removed or rearranged. The fastest way to do this is to follow the last part assembly as it moves down the line. For each station through which the assembly moves, the bins and tools are removed and replaced one at a time with the bins and tools for the upcoming product. So as the cell is being set up to assemble the new product, an operator has already started building the first unit of the new product (see Figure 7.11). The horizontal presentation of the parts and tools once again makes for a simple, standard way to change over the cell. The first unit follows the last unit of the previous product, making for a quick and efficient product changeover.

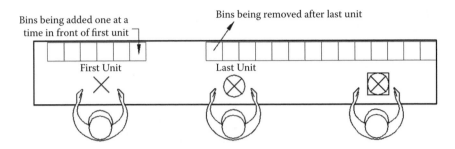

FIGURE 7.11
Quick bin change from one product to another.

In the next chapter, we will review cell design. The layout of the entire cell can vary in shape and length, depending on the processes and facility. Converting an existing cell to a self-balancing line may seem simple, but there are many things to consider during the conversion process. The layout of your cell can make or break whether the line actually achieves continuous flow.

8

Cell Designs

When you self-balance a cell, the layout of the cell is critical to flow and for connecting the work of the operators in the cell. A layout that works with a level-loaded line will not necessarily work for self-balancing; some adjustments, and often a complete new layout, may be required. This chapter will cover layout examples, design considerations, and operator-station setup within the cell.

CELL DESIGN CONSIDERATIONS

There are three primary considerations that must be addressed when designing a self-balancing cell: safety, walking distance, and part-travel distance.

Safety

A self-balancing cell is a dynamic workplace. The operators are moving up and down the cell, handing off the units—all day long—and the likelihood of any potential safety problems must be minimized.

There must be no obstructions to operators as they move through the cell. The floor must be clear of any trip hazards, and nothing should protrude from the work surfaces.

The height of the work must be adjusted for a standing operator. Ideally, for benchwork, the height of the workbench will be adjustable by pressing a button or a foot pedal to accommodate heights of various operators. Adjustable benches are more expensive but are not always necessary: operators are moving throughout the cell or other workstations, and do not remain at the same workstation all day as with a traditional level-loaded cell; this variety in movement allows operators to temporarily work at a station that may not be set ideally for their height. However, if an operator, through the division of work, is at a station greater than 50% of the operator's time in the cell, the station should be adjustable to best accommodate her height.

Walking Distance

In a self-balancing cell, the entire length of the cell is walked every takt time by the cumulative number of operators in the cell (i.e., if the cell had one operator, he would walk the entire line; two operators will each walk approximately half the line, etc.). There may be some shortcuts across a cell, but generally, the amount of walking that takes place in the cell will greatly influence the productivity of the cell. The waste of excessive movement can make it unreasonable to expect the operators to maintain a pace able to accommodate takt time. Your mantra when designing a self-balancing cell should be "compress, compress, and compress." Every inch smaller that you can make cell is waste (excessive motion) that is eliminated every takt time.

Part-Travel Distance

The distance that parts travel also is a consideration when designing a cell. If the parts do not leave the cell for any off-line processing, then the distance of part travel will be the same as the distance of operator travel. Off-line processing, then, should be arranged as close to the frontline as practical to reduce the distance that parts travel to the frontline. Similarly, the close proximity of backbench, batch, and any off-line processing also will reduce any operator travel to deliver the parts or man a backbench operation. It also will make the visual management of the inventory easier for the operators in the cell.

CELL LAYOUTS

When designing a cell, make the adage "form follows function" your overriding guideline. Often the constraints of the facility in which a cell operates will prevent an optimal design. Nonetheless, always first design a *greenfield* or ideal-state layout. If the ideal-state layout cannot be installed because of facility constraints, the least compromised version should then be installed.

If there are little to no constraints in the facility, some basic cell designs are applicable. These common concepts also can be used for more complex cell designs.

Basic Cell Shapes

The basic cell shapes are straight line and U-shaped (see Figure 8.1).

A *straight line cell* is actually a small line, and it is easy to fit into most facilities. Operators' positions in the cell are clear and easy to maintain.

A *U-shaped cell* is probably the most familiar cell design. The line-of-sight communication of the operators in the cell is ideal. Cross-training and support for the operators is concentrated in the cell and walking upstream is reduced because the operators can take shortcuts across the cell (see Figure 8.2).

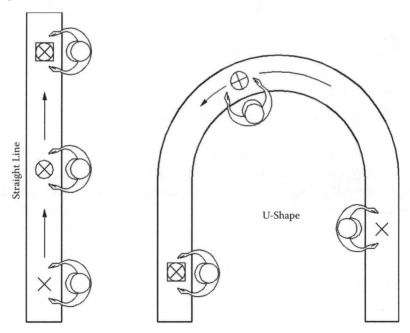

FIGURE 8.1
Straight line and U-shaped cells.

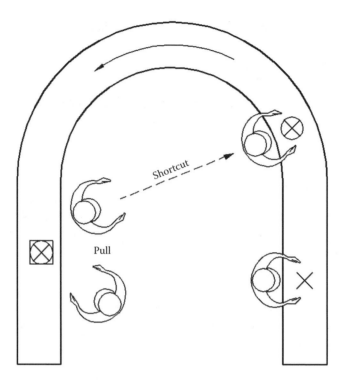

FIGURE 8.2
Walking across a U-shaped cell to pull.

Cells with Off-Line Processing

More complex cell designs require some creativity to maintain flow within the cell while minimizing part and operator travel.

Cells with backbench processes should be designed so that the frontline cell aligns with the backbench directly behind it. A straight line cell works especially well if there is more than one backbench process. Placing backbench processes in a U-shaped cell can interfere with some of the benefits of the U-shaped design; the line-of-sight communication and the shortcut routes get blocked. The cell also may have to be wider to accommodate the backbench stations (see Figures 8.3 and 8.4).

If the off-line process is a shared resource with another cell, the backbench stations can be shared between two straight line cells (see Figure 8.5).

In a U-shaped cell, the best place to set up the off-line processing is at the "turn" in the U. If there are two off-line process steps, the turn in the U can be used twice (see Figure 8.6).

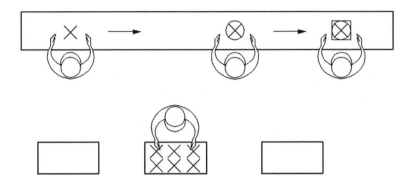

FIGURE 8.3
Backbench stations directly behind a straight line cell.

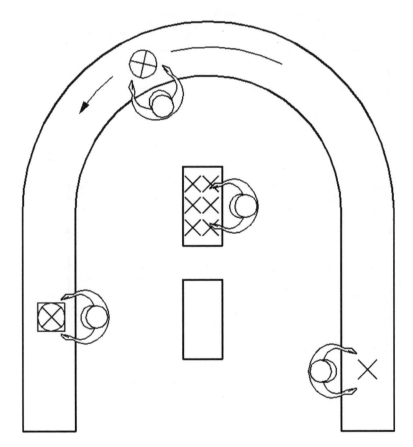

FIGURE 8.4
Widened U-shaped cell to accommodate backbench stations.

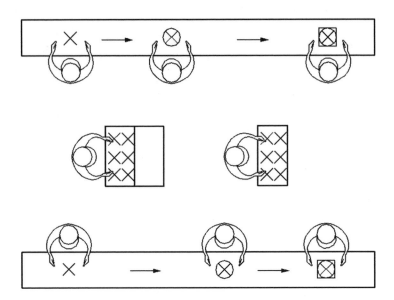

FIGURE 8.5
Straight-lined cells sharing backbench stations.

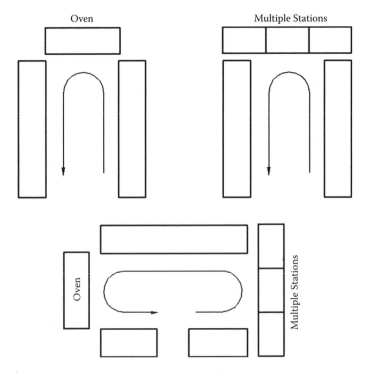

FIGURE 8.6
Backbench stations located at the turn in U-shaped cells.

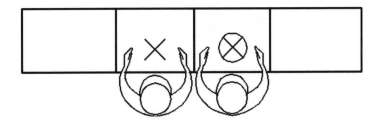

FIGURE 8.7
Minimum separation between two stations.

STATION SETUP

Right-Sized Stations

A station should be sized as small as is practical to do the work and not interfere with other operators. This means that some stations can be as narrow as the width of an operator with the operator's elbows extended to the side.

When setting up workstations, it is common to purchase benches in standard widths. The most common widths in North America are 5- and 6-foot-wide benches. These widths are usually much too wide for a workstation, and, if they were used in a cell, the overall length would be much longer than necessary as well, making the cell's footprint and walking distance oversized and wasteful.

To right-size the workstation means to design the optimal layout of the station and then build the bench to match the design. Remember, form follows function. Depending on the size of the unit being built, most work can be done on a bench no wider than 3 feet.

Vertical Space

In order to right-size a workbench, limit what is on the work surface to only that which is absolutely necessary for the operator to perform at the station. Creative use of the vertical space should be used for everything else. The work surface is "prime real estate" and should not be cluttered with anything that will increase the width of the station.

Vertical storage can be used for common items, including:

- Computers, monitors, keyboards
- Tool storage

- Components
- Consumables
- Waste or scrap material

Standing Height

Because the cells in a self-balanced line are standing/walking cells, all of the work performed and the work surfaces must be set up for standing height. The bench height can be determined by setting the height at the highest level at which the shortest operator can comfortably and safely work. If the benches can easily be adjusted, the maximum height should fit for the tallest operator. If the benches are not adjustable or additional height adjustments are needed, simple, creative, and low-cost solutions can be used. For example, a sturdy tote turned upside down can be used as an optional-height work surface. But whatever means is used to achieve the proper standing height must be stable—raised benches and equipment can be top-heavy, and the leverage created by working at those heights can add to any instability.

Standing Surface

The surface that the operators stand on is an important part of the cell design. Standing and walking all day on concrete, linoleum, or other hard surfaces is very tiring to an operator. Use one of the many types of anti-fatigue mats wherever possible.

Modular antifatigue mats can be easily moved and reconfigured if the cell layout changes. Large mats are less flexible and quite heavy to move around.

The edge of the mats should be tapered to the floor to reduce the trip hazard when walking on and off of the mat. The mat also should extend under the workbench far enough so that an operator's feet do not hang off the edge when standing up against the bench.

If the switch to self-balancing requires operators who have predominantly worked sitting down, to now work standing up, operator safety

Quick Adjustments

Any time spent adjusting a workstation should be a small fraction of the takt time. Long setup times will cut into the productivity gains of self-balancing.

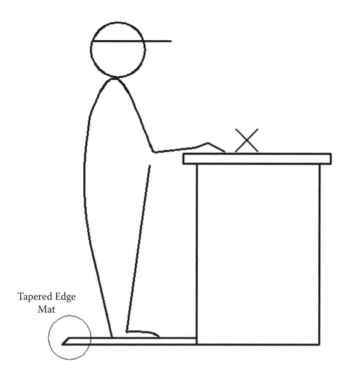

FIGURE 8.8
Antifatigue mat with tapered edge.

must be managed: For operators who have not worked on their feet before, it can take a few weeks to get "in shape" and become comfortable working on their feet all day. During this adjustment period, it is important to support the well-being of the operators and make sure they are not getting too fatigued. Adequate breaks and rotation should be provided. If there is room in the cell or near the cell, a resting chair can be provided for an operator to take a mini-break.

The importance of a good cell design cannot be overemphasized. A compact, well-thought-out layout means that your transition to self-balancing will avoid the designed-in waste of excessive walking.

Conclusion

You now have the tools you need to create continuous flow in your factory. For simple applications, it is a no-brainer to implement self-balancing.

For the more complex processes and assembly lines, you can also achieve true continuous flow. Although it has been easy to assign small, discrete tasks to a worker, it will initially be a bigger responsibility to link up all of the discrete tasks to create flow all the way through. It is what we have been striving for, for a long time, and it is well worth it.

Your operators are now working and collaborating as a team, possibly for the first time. You now need only manage the team. The days of trying to optimize individual performance and efficiency are over. It drove the wrong behavior and was counter to creating flow. The team can continually improve the way they work together. The ways that they experience their job, coworkers, and products will transform. No longer will they have an individual silo mentality, performing the exact same work over and over.

The benefits of self-balancing are many. Some will be realized over time, as you gain a deeper understanding of the process. Besides the productivity, WIP, and lead time reduction, one of the greatest benefits of self-balancing is having a process that will only work with a standard work sequence. While establishing and maintaining a standard work sequence could have proved to be a challenge in the past, self-balancing lines will not run with each and every operator performing the work in the exact same order, shift to shift, day after day, starting Day One. This benefit cannot be overstated.

The more you apply the self-balancing process, the more you begin to master it and see where *else* it applies. Because self-balancing has such a bias for flow, anything else really begins to stand out as wasteful. Self-balancing now is the new standard for creating continuous flow. Any process, whether it is manufacturing, business processes, product development, transactional, or more, may be a candidate for self-balancing. If you observe queues, have a standard, progressive sequence for applying the work, and it is out of balance, you can see how self-balancing applies.

In the spirit of continuous improvement, self-balancing gets us closer to a perfect process, achieving true continuous flow. While balancing a line through level loading will always have its applications, and a one-person bay-build is simple and always balanced, self-balancing can now be another choice for creating flow. Give it a try.

Index